The Urban Planning Imagination

For 苗田
and
Yulia Elizabeth Phelps

The Urban Planning Imagination

A Critical International Introduction

NICHOLAS A. PHELPS

polity

First published in 2021 by Polity Press

Polity Press
65 Bridge Street
Cambridge CB2 1UR, UK

Polity Press
101 Station Landing
Suite 300
Medford, MA 02155, USA

ISBN-13: 978-1-5095-2624-6
ISBN-13: 978-1-5095-2625-3(pb)

A catalogue record for this book is available from the British Library.

Typeset in 11 on 13pt Scala
by Fakenham Prepress Solutions, Fakenham, Norfolk NR21 8NL
Printed and bound in Great Britain by Short Run Press

For further information on Polity, visit our website:
politybooks.com

Contents

List of figures and tables

Figures

Tables

Preface

A good number of friends and colleagues have been implicated in yet another of my book-writing enterprises and this preface can only touch on some of them.

I must say at the outset that I am extremely grateful to Jonathan Skerrett and Karina Jákupsdóttir of Polity Press. I thank Karina for keeping me on the straight and narrow and Jonathan for the invitation to write this book and for his insightful comments and persistence with my attempts to put together a cohesive framework for it. I thank Fiona Sewell for her cleaning up of my manuscript text. I am greatly indebted to readers of the original book proposal and reviewers of the first draft manuscript for their constructive comments. I have not been able to respond to all their suggestions, but I hope they see something of themselves in this book. Mark Tewdwr-Jones was a source of both enthusiasm and ideas at the early stages of writing and several of the chapters bear the marks of his thoughts. I thank Miles Irving for preparing the figures contained in the book.

Many of the ideas in this book were developed while I taught the subject of International Planning (which then became the two subjects, Critical Debates in International Planning and Comparative Planning Systems and Cultures) over the course of eleven years at the Bartlett School of Planning, University College London. These subjects formed a core part of the MSc in International Planning offered there and I am grateful to Claire Colomb, Nick Gallent, Nikos Karadamitriou, Claudio De Magalhaes and Susan Moore among others for their comradeship, and to the postgraduate students who passed through the programme for their participation and enthusiasm.

The book also contains more than a hint of Australia and Australian urban conditions in it. Some of the ideas were tested out on first-year

Bachelor of Design students at the University of Melbourne without any apparent side effects. I thank senior tutor Dejan Malenic for his sterling efforts in helping with the day-to-day running of the subject. I am also grateful to academic colleagues at the University of Melbourne: Michele Acuto, Judy Bush, Stephanie Butcher, Patrick Cobbinah, Brendan Gleeson, Anna Hurlimann, Crystal Legacy, Alan March, David Nichols and John Stone among others for their unsuspectingly steering me in the direction of several new sources of planning inspiration. Elsewhere in Australia, Robert Freestone at the University of New South Wales and Paul Maginn at the University of Western Australia have been valuable sounding boards for my evolving urban planning research ideas. Jago Dodson at RMIT has helped join up a few of many Australian dots for me.

Further afield, Roger Keil at York University, Canada, and Fulong Wu at University College London continue to be sources of sound advice, and Dave Valler at Oxford Brookes University a great friend, co-researcher and co-author on urban planning matters. As visiting scholars to the University of Melbourne, Martin Arias (Universidad Catolica del Norte), Ben Clifford (University College London) and Andy Wood (University of Kentucky, Lexington) helped more than they will have realized.

The most important person helped just by being twice her amazing self.

Nicholas A. Phelps
Melbourne
Australia

1 Introduction: what is planning?

Introduction

Modern urban planning has been defined in many ways that shed light on this multifaceted activity. Kunzmann (2005: 236) suggests that urban planning is 'the guidance of the spatial development of a settlement'. The Commission of European Communities (CEC, 1997: 24) defines spatial planning as 'the methods used ... to influence the future distribution of activities in space ... to co-ordinate the spatial impact of other sectoral policies ... and to regulate the conversion of land and property uses'. To quote Magnusson (2011: 131), 'to plan the city is to rationalize our activities in relation to one another within a confined space, but it is also to think of how that space is to be reshaped as a sustainably habitable, productive, comfortable and congenial place'. Just as there is a sociological imagination that exceeds questions of 'personal ingenuity and private wealth' (Mills, 1959: 10), so there has been and should continue to be an urban planning imagination at work in the way we settle the earth. Adapting dictionary definitions, that urban planning imagination might be defined as the faculty for forming ideas, images or concepts relevant to the task of city building where these need not be entirely new but instead are the products of an historical stream and geographical diversity of ideas, images and concepts. The 'huge amount of energy expended on "planning" as demonstrated by the multiple types of plans at all levels' (ESPON, 2018: 76) suggests that urban planning is an increasingly pervasive and indispensable activity – one that is a geohistorical stream of thoughtful and practical acts that carry valuable wisdom of what works, what doesn't, what could be desirable and what is not.

'Planning is both anticipatory and reactive' (Levy, 2016: 6) and both aspects are found across several different styles elaborated in a recent review of European planning systems (ESPON, 2018). Regulative planning is legally binding while framework-setting planning typically involves non-binding policies. Strategic planning typically provides non-binding indicative reference points regarding future development, and visionary planning sets out agendas for desired futures in the form of normative principles. Planning's regulatory aspect leads strongly in the direction of rules or codes regarding the use of land that shape 'how places perform – socially, environmentally, and economically' (Talen, 2011: 11). Codes and standards have long existed for building construction, the layout of entire settlements and streets and key spaces within them to ensure the safety and health of urban populations, but they are also carriers of societal values (Ben-Joseph, 2012). Rules shape places imperfectly and elicit great creativity directed at evading or distorting those same rules. Yet the drift in many parts of the world has been inexorably towards more regulation, despite the largely counterfactual question of whether non-planning, less regulation or zoning would be worse (Banham et al., 1969; Siegan, 1970). Of course, in the absence of the sort of imagination associated more with strategic and visionary styles of urban planning, the urban experience may become no more than a surrender to codes and their unintended and unanticipated consequences.

None of the definitions of urban planning and its associated imagination noted above are prescriptive about *who* is doing urban planning, since, as Wildavsky (1973: 129) noted, 'planning must not be confused with the existence of a formal plan, people called planners, or an institution'. In this sense, the attempt to distinguish urban planning from non-planning – perhaps 'the market' – is futile: the two are inseparable.[1] Urban planning is pervasive, as John Friedmann (1987: 25) noted when defining it as part of the public domain and as 'a social and political process in which many actors, representing many different interests, participate in a refined division of labour'. It is to be found 'at the very centre of the complex mass of technology, politics, culture and economy that creates our urban society and its physical presence' (Rydin, 2011: 1–2). Thus, 'many of the so-called market forces that the planning system takes as given are in fact caused by public policies to which individuals and businesses respond' (OECD, 2017b: 17). The outcomes of planning past and present are made plain in the appearance of cities and patterns of settlement.

The pervasiveness of urban planning leads me to take a broad view of the range of activities, actors and associated ideas and methods that constitute it. Far from being empty, the injunction to plan concerns the unavoidable need for purposive thought and action as a way of responding to our being in the world; we all act, make plans for ourselves, and are acted upon by the plans of others. If, as philosopher Edward Casey (1997: ix) observes, 'to be at all ... is to be in some kind of place', then the urban places that we have imagined, designed, planned and made for ourselves are the very expression of our being in the world. They express the uneasy tension between our own sense of self and our peaceful and productive coexistence with others. In this view, 'plan making was an established art long before even a modest portion of human settlements could be regarded as urban' (Silver, 2018: 11). Moreover, 'place is as pervasive and important as language: we are place-makers and users, as we are language-makers and users' (Sack, 2003: 4). In our inability as humans to accept reality (nature as it is/was), we engage in (urban) planning as a purposeful, future-oriented act of imagination. Regardless of whether it comes on a grand scale or in increments, at its best, urban planning is concerned with shaping good/better places, though it can just as easily – with lack of awareness and thought – lead to the production of poorer, bad or downright evil places (Sack, 2003).

We live in an urban age – an age where the majority of the world's population live in officially defined urban areas. The United Nations (UN) has estimated that two-thirds of the world's population will live in cities by 2050 (United Nations, 2014). I use the term 'urban planning' rather than 'spatial planning' not to deny either the quantitative significance of vast expanses of rural and semi-rural lands that lie outside the world's formally defined urban areas, or indeed to deny their relationships to cities, but to signal instead the renewed significance of developing the planning imagination in this urban age. The need for us to plan settlements in ways which are sensitive to the vast natural hinterlands from which they have been carved is more pressing than ever in an age in which urbanization is associated with enormous consumption of natural resources, the production of waste and greenhouse gas emissions (Camaren and Swilling, 2012). The need to do so in ways which recognize and leverage the historical and geographical interrelationships that inhere in thought and action distributed across an increasing array of actors has never been more pressing. This is the sense in which I speak of the urban planning imagination.

The urban planning imagination

The urban planning imagination is ever more distributed across a range of actors with differing geohistorical sensibilities. It is this that ensures that consideration of urban planning's contributions and failures should adopt vantage points well outside those of Western Europe and North America. The way in which we think about urban planning, as professionals, educators, politicians, civic activists, business and association leaders and citizens, should perhaps be forgiving of urban planning's inherent limitations but re-enchanted by its impressive and growing stock of knowledge, ideas and methods and the sense of possibility it carries with it. To plan – as to err – is human.

Urban planning has a geohistory and imagination that far precede planning as a modern profession, and range from indigenous Australians' complex relationships to land to the cities of Mesopotamia, Imperial China, Athens and Rome and those of Latin and Meso-American civilizations, through to the cities built in the Renaissance in Europe and in, for example, the Philippines, Peru and Mexico under the Laws of the Indies – where in each case significant financial and human resources were devoted to city planning and building (Hein, 2018). Indeed, 'many of these earlier interventions are still visible ... They continue to shape practice in multiple ways, through governance structures or planning cultures, through inherent path-dependencies of institutions or laws and regulations, as formal references, or frameworks for design, transformation, and preservation' (Hein, 2018: 2).

Who plans?

In this book I argue that if urban planning is part of 'a refined division of labour' (Friedmann, 1987) then it has become a more complex and distributed set of practices as the division of labour in society continues to evolve. The innumerable acts, the substantive concerns, wisdom and methods, and the most inspiring and powerful historical and geographical references for urban planning are apparent across a diversity of actors that I simplify here as citizens (individuals and individual households), clubs (corporations, civic associations, environmental groups etc.) and states (new, old, unitary, federal, liberal market, developmental etc.). The interest, influence and power to shape urban development outcomes are distributed very unevenly across these actors, with states and their planning pervasive but less powerful in certain respects than is often appreciated (McGlynn, 1993). The urban planning imagination speaks to and operates in and through 'a patchwork of private, club, and public realms that both cohere and fragment the city' (Webster, 2002: 409).

It may be particularly important to recognize the diversity of urban planners and urban planning practices found in and across citizens, clubs and states in the modern era, when it is all too easy to reduce urban planning – its imagination, its substantive concerns, wisdom and methods – to the institutionalized statutory urban planning of the global north in the past 150 years or so. To be sure, the institutions of statutory planning provide a store of wisdom: 'precedent does offer access to a rich archive of prior human experience and creativity' (Hoch, 2019: 99). However, much of the emotional intelligence that Hoch (2019) directs us to and which can provide new, practical, urban planning wisdom may rest with citizen and club actors to be mobilized in productive mixes between state, citizen and club, as I emphasize at points throughout the book.

Instead, then, the strengths and imagination of urban planning are to be sought in the increasingly dispersed nature of innumerable, more or less reflexive, acts by citizens, and in the name of clubs and states across sweeps of time and space that collectively describe the making of cities. If learning itself remains the most valuable resource people possess to prepare for the future (Hoch, 2019: 3), the future of the urban planning imagination will need to be open to the complex possible mixes or combinations of, or experiments among, citizens, clubs and states found in different parts of the world at different times. The positive contributions to city making of some of these mixes may seem unlikely, but we should suspend any prejudices we may harbour here regarding the essential properties or rationalities of citizen, club or state planning actors if we are to continue to offer broadly popular and tractable, if temporary, solutions to the unending stream of challenges that attend city making.

History and the urban planning imagination

An historical perspective on cities and urban planning is needed since, as Patrick Geddes (1904: 107) argued, 'a city is more than a place in space, it is a drama in time'. The securing of shelter from the elements and the mobilization of 'things to hand' are central to the human condition of *becoming*. As with the sociological imagination (Mills, 1959), the urban planning imagination must firmly locate itself within the stream of individual, club and state actions by which our cities are built. Acts of urban planning emerge as something ordinary in their immediacy and yet extraordinary in their longer-term effects. An historical sensibility – a reflexive sensitivity to the temporality of the city and urban planning itself – is vitally important to understanding the becoming of cities. Conservation of the natural and built environment is an important substantive concern of urban planning. History is important to excavating and understanding the failed or lost potentials of cities and associated urban planning imaginations. However, the backwards look can never be the majority part of urban planning, let alone its entirety. This is why I choose to define urban planning as an imaginative, future-oriented act even as it is cognizant of the past.

Urban planning is characterized by significant *continuity* as a result of particular, durable, administrative and legal traditions that inhere within societal cultures and, more recently, in the statutory basis of national and consequently local planning systems, as I discuss in chapter 6. Continuity is a product of the habits and conventions adopted and acquired by generations of planners – whether individuals, corporations or states – and which become fossilized in policies, plans and meanings and values attached to particular sites in what we understand as distinct urban planning cultures. At the same time, it is apparent that urban planning has been the subject of significant *change*. Urban planning activities and processes are subject to multiple temporalities or rhythms (Abram, 2014) from the long term of scenarios, to the mid-term of forward or strategic spatial planning, to the short term of decisions on individual development proposals. Indeed, there is a sense in which *change may be the only constant of urban planning*. Histories reveal both important changes in the substantive concerns of urban planning over time in single places and historical slippages in these same concerns from one place to another. It should be clear, then, that the history of urban planning is not linear – heading inexorably in the direction of 'progress'. History repeats itself in terms of how the substantive concerns of urban planning come into and go out of view, and how the wisdom associated with urban planning is valued or undervalued.

Geography and the urban planning imagination

It is in place making and shaping that the definition of urban planning I have in mind has an inherently geographical aspect to it. The planning imagination must be a geographical one in its attention to the uniqueness of places. Geddes (1904) considered geographical method as fundamental to the comprehensive understanding of cities and their urban planning, drawing on geographical notions of the unity and coherence of places or regions, not least because 'it takes the whole region to make the city' (Geddes, 1904: 106). Seeing the city in these terms has been part of a modern planning tradition of the past 150 years and it 'often seems a messy, conflict-ridden and threatening enterprise because it seeks to "integrate", to connect, different areas of knowledge and practice around a place-focus' (Healey, 2007: 13).

It is clear both historically and in the present that our cities have never been, and can never be, entirely closed or disconnected places. They are shot through with physical, virtual and remembered references, relations and connections to other places. It is vital, then, for the urban planning imagination to bring to bear a perspective on the seemingly general or universal nature of our urban existence. This sense of the partial convergence on more or less universal elements of urban planning is familiar to us in the shorthand term 'globalization'.

If urban planning is an activity involving the shaping of places, then it is an act of imagination that must seek to reconcile these two geographical perspectives. It is a thoughtful activity in and through which what Doreen Massey (1989) termed a relational or global sense of place might be mobilized. Thus, in chapter 2 I will elaborate how a geographical perspective reveals the 'betweenness of place' (Entrikin, 1991) as both unique and bounded but also permeated by more or less common (cultural, economic, social, environmental) processes and relations. This geographical perspective reveals both the distinctiveness of different planning systems (chapter 6) and some of the elements of convergence and exchange among them (chapter 7).

Urban planning's enduring appeal

Urban planning emerges as an activity that has adapted to changing societal needs and desires, retaining an element of imagination while acting to bind a variety of actors and their interests in efforts to address the substantive challenges associated with human settlement around the globe.

The urban planning imagination's geohistorical sensibilities make it a particularly powerful and integrative means for solving the complex problems of city building, since these reveal themselves as ones of (spatial) interdependence and indivisibility and (historical) uncertainty and irreversibility (Hopkins, 2001). The urban planning imagination – as something distributed across citizens, clubs and states – emerges as pervasive but more suitably modest (Hoch, 2019: 48) than it has at times been presented as being in statutory practice and university training issuing from the global north.

My celebration of urban planning is not one that rests on the thought that urban planning is somehow an unconstrained act of imagination; it is not. Rather, as I note in chapter 6, urban planning systems and cultures are nested within broader institutional and cultural frames while being an indispensable part of, or foil to, them. As Magnusson (2011: 132) observes, 'Planning has always been a way of rationalizing politics by rendering it governable.' Indeed, urban planning has had 'greatness thrust upon it' at various junctures. These include, for example, the aftermath of war in the United Kingdom (UK), when urban planning was briefly *the* means by which the modernization of society was to be achieved (Hall and Tewdwr-Jones, 2020), and the present, with urban planning emerging as the most suitable arena in which to address the effects of climate change and search for sustainable development (Davoudi et al., 2009). At its best, urban planning continues to manifest something of society's collective conscience in connection with what Rittel and Webber (1973) explain are 'wicked' problems.

It has been said that urban planning is a dialectical process (Gleeson and Low, 2000) whose tensions are reconciled in *moments* in time and place. Often a particular visual (map, diagram, sketch), technique (forecasting, overlays), method (scenario building, collaborative or communicative processes) or principle (sustainability) captures the imagination. At these moments the incredible global mobility of urban planning thought and practice becomes visible.

It is conceivable in a world now considerably sped up that those moments in which urban planning gains purchase will be too fleeting to be meaningful. Yet, in other respects, the speed of change makes urban planning an even more important enterprise in the present age, though one in need of rethinking as a joint exercise drawing sustenance from the distributed and pervasive nature of urban planning itself: drawing strength and imagination from the substantive concerns, wisdom and methods found across a range of actors.

The structure of the book

The sketch of the urban planning imagination presented here is one of variations on reasonably common themes: one in which there are not only contrasts between cities and nations but also common pressures upon them, much exchanging of ideas and no small measure of similar – though by no means identical – policies and practices developed and deployed. These contrasts and common-alities are visible across places – cities and nations – and across time in the same place. Geographical and historical perspectives are essential to the study and appreciation of urban planning. To the extent that any individual, white, global-north-rooted male can obtain the knowledge and muster the powers of expression required for a critical international introduction, my short discussion of the urban planning imagination cannot be anything other than partial.

In the following chapter, I begin by offering a partial answer to the question 'who plans?'. I define urban planning actors in terms of citizens, clubs and states before going on to note the ever more mixed properties of the urban planning imagination apparent within and across these sets of actors. I then depict the close links that planning as a discipline has with the study of history and geography. The urban planning imagination is something which unites historical and geographical sensibilities and animates them with a sense of normative purpose towards the shaping of better/ good places. It is a body of thought and practice that is uniquely and purposively integrative and synoptic in its aspirations. In this it has rarely been completely successful, for the shaping of better/good places is a task and work necessarily never quite finished. As one of the most criticized and insecure of disciplines, planning can hardly escape a sense of its own fallibility. If this is one lesson painfully learned, it is one that other disciplines might do well to incorporate a little of.

Chapter 3 sets out some of the substantive concerns – shelter, health, mobility, sustainability and economy – that urban planning has had to address. These appear as enduring issues for urban planning thought and practice to deal with. Nevertheless, the precise nature and severity of individual issues and their pecking order continue to alter over time, shaped by circumstance.

Chapter 4 reminds us why urban planning has a value. It has a continuing value as a stock of inter-generational knowledge dating back even to humankind's first settled relationships to nature. There is more than a sense of lessons not learned, things forgotten, wheels being reinvented and history repeating itself here. However, the stock of urban planning's wisdom continues to grow in ways that will help us with the challenges of place making that lie ahead.

Chapter 5 discusses some of the methods associated with urban planning. Something of the enormous imagination of urban planning is again showcased here, from the details of techniques used in particular instances, to the long-term 'informed speculation' associated with 'what-if' scenarios, to the ways in which we can seek to mobilize the intimate local knowledge and intensely felt needs and desires of citizens. This multiplicity of methods is further evidence of the integrative and synoptic potential of urban planning thought and practice.

Chapter 6 seeks to illustrate one aspect of urban planning's geographical sensibility by drawing attention to the variety of different systems and cultures that exist. Urban planning thought and practice reside within and take their cues from the broader culture and institutional arrangements of societies. The variety of urban planning systems and cultures is itself a stock of accumulated knowledge and expertise, the surface of which has barely begun to be scratched in the extant academic and practice literature. Its full significance is exposed when one realizes that it is the driver of the sorts of international exchanges that I discuss in chapter 7.

Urban planning thought and practice have long been exchanged. City plans have been copied, ideas for unrealized cities have provided inspiration, and particular policies or techniques of planning have been adopted widely. To add to this, apparently similar urban forms and principles of urban planning have developed in synchrony in different parts of the world. What are we to make of this? To me, this signals the power of the urban planning imagination. Urban planning has appeal because it is needed. It has seduced and continues to seduce. How we reflect on and mobilize this power will be important to urban planning in an urban age.

I conclude in chapter 8 by considering the future of urban planning in the present age. This is a future of the mixing of actors, the knowledge and wisdom they bring to substantive challenges, and the methods they make use of when drawing from historical and geographical vantage points. It will need to be a progressive mix with purpose rather than one that produces lowest-common-denominator outcomes, an incompatible pick and mix, or partial and exclusive combinations of citizen, club and state urban planning imaginaries.

2 Imagination: what is planning's spirit and purpose?

Introduction

The planning imagination has been at work in the way we have built cities, but what kind of imagination has been apparent, across which actors, to what ends, and what might that imagination look like in the future? The urban planning imagination is not the exclusive property of one set of actors. Urban planning's spirit and purpose (Bruton, 1984) will be found in new and productive mixes of imaginations regarding present and future urban planning challenges.

We should not confuse the future orientation of the urban planning imagination with a loss of historical perspective, for planning needs to 'broaden its preoccupation with space, and to take consideration of time' (Wilson, 2009: 232). History plays into the present and future of urban planning in complex, non-linear ways in which the imaginative aspects of planning make it 'a kind of compact between now and the future' (Abram and Weszkalays, 2011: 8). Geography is part art, part science (Entrikin, 1991). Likewise, 'making plans for places is more craft than science' (Hoch, 2019: 4). Indeed, 'plans are unique forms of public policy. Both art and science, they embody a vision of the future for which there is no proof' (Hanson, 2017: 262). We should not confuse the ordering of settlement space with the inexorable contiguous growth of a city, since the decline and abandonment of cities has a long history. Relational senses of place have flourished within which the city can be understood not merely as a bounded place but also as a node within networks of places or a nexus of flows or virtual connections. These sensibilities reveal the uniqueness of places and the commonalities produced through connection.

This geohistorical sensibility describes urban planning's value to societies, capitalist or otherwise. The imagination of urban planning exceeds that of history or geography in that it has a normative aspect – the desire to produce better (or good) urban places. The urban planning imagination is a force for integration and inclusion in a world in which we can all too easily grow further apart. We would have to invent it if it did not already exist.

Who plans?

By way of simplifying the story, I refer to three sets of urban planning actors: citizens (as individuals or households), clubs (e.g. multi-national enterprises (MNEs), associations of mutual interest, private enterprises) and modern nation states (their local governments and the international interstate system). Recognition of the different actors central to urban planning throughout history implies nothing essential of the motives, the substantive interests or expertise, the wisdom, the geohistorical sensibilities, the sophistication of the methods involved, or the outcomes achieved. The motivations that lie behind acts of urban planning can be obscure both at the time and afterwards and are rightly open to scrutiny, debate, argument, objection and protest. It should also be clear that the urban planning of clubs and states is hardly any less varied in its motives and outcomes than that of myriad citizens. There is as much variety *within* each category of urban planning actor as there is *between* citizens, clubs and states. There is yet more variety to uncover in the 'experimental' overlaps in the imagination, substantive interests, wisdom and methods of actors – as depicted in figure 2.1. Much of this experimental variety has yet to be recognized, let alone unlocked, as I discuss further in chapter 8.

Our settlements are the collective creations of citizens, clubs and states. They are triumphs (Glaeser, 2011) but they are not free from the significant conflicts and imperfections I discuss in chapter 4. Cities are made in our own image and are the physical expressions of both our better and our darker nature. The damage done to the indigenous peoples of Australia and the Torres Straits Islands is testimony to the brutal power of urban planning to deny ancient ways of being-in-place in the process of colonial settlement (Jackson et al., 2017). Elsewhere, in China, planning has been more positively connected to the preservation of ancient *urban* civilization (Morris, 1994).

One set of these actors can predominate in the planning of cities. The earliest cities of Mesopotamia might be considered

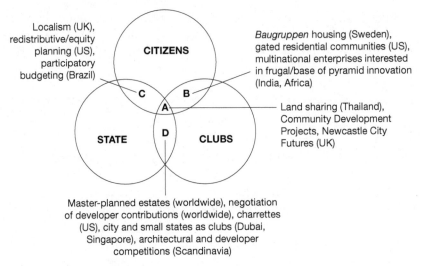

Localism (UK), redistributive/equity planning (US), participatory budgeting (Brazil)

CITIZENS

Baugruppen housing (Sweden), gated residential communities (US), multinational enterprises interested in frugal/base of pyramid innovation (India, Africa)

C   B

A

STATE   D   CLUBS

Land sharing (Thailand), Community Development Projects, Newcastle City Futures (UK)

Master-planned estates (worldwide), negotiation of developer contributions (worldwide), charrettes (US), city and small states as clubs (Dubai, Singapore), architectural and developer competitions (Scandinavia)

Figure 2.1 Urban planning actors and mixes of actors

concentrations of many citizens. In Europe the city emerged as a club – a municipal corporation – to shield citizens from the powers monopolized by new nation states (Frug, 2000). Cities continue to emerge within the nation-state system in many privately developed new town clubs across the global north and south. Finally, cities have manifested as states. Minus some of the civic ideals, the ancient city states of Athens and Rome have their modern-day counterpart in Singapore.

The balance of these actors in the making of individual cities has varied over time. The historical evidence of a mix of actors notwithstanding, I suggest that new combinations of actors may become the defining feature of urban planning as it is emerging, and we will need to better understand the possibilities for these combinations rather than be led by prejudice regarding the motives or capacities of citizens, clubs or states.

Citizens

Some form of spatial awareness, organization and planning has been vital to survival, since to dwell in a place is to have regard for 'things to hand' (Heidegger, 2010). This is the existentialist sense in which planning precedes rationality rather than being guided by it (Hoch, 2019). These individualistic urban planning tendencies are an irrepressible aspect of human nature. The mass expression of need for shelter across the global south may instead be latent in the highly regulated planning systems of the global north. Regardless, 'citizens correctly assume that they know something about planning without having studied it formally' (Levy, 2016: 94).

On the one hand, then, the continuing concern in the global north to define and protect the planning profession as primarily a statutory activity seems out of place when marginalized peoples – in the global north and south – take into their own hands the task of organizing their housing and, by extension, their immediate neighbourhoods and cities (Miraftab, 2009). The lay knowledge and emotional intelligence of citizens are things we might reasonably wish to better incorporate into the planning of our cities (Hayden, 1997; Hoch, 2019). Yet it is as well to remember that 'Citizens, like elites, can be misguided and self-serving' (Fainstein, 2010: 32), whether in NIMBY ('not in my back yard') protests against development or in opportunistic capturing of the spoils of urban development. In the absence of political will and bureaucratic resources to address vexed issues of compensation and betterment (chapter 4), local government planners can appear powerless to shape equitable outcomes from the development of cities.

It is important to recognize citizens as a distinct set of actors in the building of our cities. As 'users', citizens have an intense interest in the city form and function even if they are perhaps the least powerful of actors shaping it (McGlynn, 1993). 'A viable notion of empowerment of the poor requires an appreciation that empowerment is functionally an individual process that deepens with time if individual efforts are consciously embedded in more collective forms of ... mutual empowerment' (Pieterse, 2008: 7–8). Indeed, citizens often participate in groups or clubs of diverse complexion and influence (Levy, 2016: 98), and it is these collective or club forms of mutual empowerment I turn to as a second category of urban planning actors.

Clubs

Access to the public realms we most associate with the city is uneven because most of these spaces generate an inclusion/exclusion problem, derived from some need to ration use or from the costs in time and money needed to access them (Webster, 2002: 398). New

York is the setting for numerous private public spaces (Kayden, 2000) and the reality is that 'most public realms serve *particular* publics and are better conceived of as *club* realms' (Webster, 2002: 398).

There is a diversity of entities planning, producing and managing urban space that could be lumped together as 'clubs'. They range from the worst forms of for-profit enterprises, to business-as-usual for-profit enterprises, to radical and socially and environmentally progressive communal experiments. There is nothing essential about these clubs in terms of the aesthetics or equity of their contributions to the built environment or the methods by which they seek to achieve their ends.

At their best, clubs present viable and innovative contributions to city building. Garden Cities and new towns across the world have been planned and delivered through corporations that provide urban services in the form of club goods. The Garden Cities of Letchworth and Welwyn in the UK were innovative and successful – developed in an orderly way over the business cycle, providing for schools, hospitals and parks without seceding from the rest of the urban system. Yet the same ideas have been part of the perpetuation of systemic inequalities conceived in apartheid South Africa (Skinner and Watson, 2018).

At its worst, club planning has detracted from cities and reveals ignoble motivations – of free-riding, cost cutting and reneging on promises. New master-planned communities developed by private for-profit corporations outside Jakarta in Indonesia, Johannesburg in South Africa, Santiago de Chile or Buenos Aires in Argentina might be efficient ways to deliver some urban services, but they are barely connected to the existing urban fabric or to one another by adequate road or rail infrastructure, and are sufficiently impermeable to be examples of the secession of socio-economically homogeneous segments of the population from their urban hosts.

Just as citizens can become empowered in clubs, so the ultimate measure of clubs as actors is whether they can find some of their better nature in a 'club of clubs' by which they can continue to make contributions to urban society.

States

Local and national states and the interstate system can be considered a third set of distinctly modern urban planning actors. Although 'the act of conscious town building ... has extended over millennia ... Town planning ... was different, resting as it did on notions of extension of public control over private interest in land and property' (Cherry, 1996: 17). Statutory urban planning – the legally prescribed processes by which plans are enabled, made, revised and subject to scrutiny, appeal and enforcement – has a power that the planning of citizens and clubs can rarely match. Although full of unintended and unanticipated consequences, urban planning by states has become indispensable to the continued functioning of capitalism (Scott and Roweis, 1977); so much so that it is statutory planning that we most associate with urban planning. Urban planning in this incarnation expanded its remit from one of ameliorating the worst excesses of citizen and club planning – as a protector of the commons and of the 'public interest' – to the point where its basis in private property rights has become obscured (Blomley, 2017).

Nation states are a recent form of societal organization, dating to the Treaty of Westphalia in 1648, but statutory urban planning itself took another two and a half centuries to become a distinct and important activity. Much of the repertoire of urban planning taught in planning schools pertains to the unique moment (Graham and Marvin, 2000) – a mere century and a half – in which states have assumed much of the regulatory and visionary burden of urban planning. Nation states 'see' from above (Scott, 2000) whether building new capital cities (such as Brazilia) or demolishing, redeveloping or extending existing cities.

However, even within this short history of statutory urban planning, different historical and geographical vantage points make it a difficult enterprise to generalize about. The early 'experimental' urban planning undertaken in the name of empires was one of brutal theft and segregation. Yet it was also invested with some nobler aspirations that have today been lost. In the improvement works undertaken in colonial cities, for example, 'the term "trust" carried with it an implied association with the public good rather than private profit making. Later ... in the twentieth century, it was supplanted by the terms "board", "corporation" or "authority", although the functions remained similar' (Home, 2013: 84).

Where once, in the global north, statutory urban planning was central to the building of new societies and commanded popular and political support, its value is now questioned. Even so, the inability of some nation states to exert authority and provide security over their territories often plays into poor urban planning outcomes. In some African nations the state has been absent from knitting together the self-building of citizens and the limited publics of club communities with any residual sense of the public interest (Parnell, 2018). Thus, familiar aspects of statutory urban planning remain on the agendas of supranational bodies such as the European Union (EU), international interstate organizations such as the UN, and innumerable non-governmental organizations (NGOs) that feed off the international interstate system.

Contrasts can be drawn among national planning systems and cultures which reveal some of the global diversity of statutory urban planning and citizen and club engagements with it (chapter 6). Equally, it can be invidious to apply labels to nations, especially when these speak with language issuing largely from the global north.

Mixes of actors

The urban planning imaginations of citizens, clubs and states continue to evolve in complex ways which generate overlaps in figure 2.1. Statutory urban planning processes remain an inescapable reference point for understanding how the urban planning imagination is shared across actors, as chapter 6 confirms. In the United States (US), the majority of urban planning in the early 1900s was by consultants. The rise of the local government planner in the US, as elsewhere in the global north, took place after the Second World War and with the extension of cities. Even so, perhaps as many as one third of professional planners in the US are consultants today (Pollock, 2009). In Australia, consultants are disproportionately accredited when compared to their equally numerous public sector counterparts (Elliott, 2018: 27). Consultants produce much of the evidence base used in planning in liberal market economies (Batey, 2018). The division of labour in which the imagination of state planners mingles with that of citizens and clubs continues to evolve. Across the global south, the numbers, training and resources of state planners mean that they struggle to exert influence on the actions of citizens or powerful club interests. Clubs have the substantive foci and often the human and monetary resources to compete with or augment the urban planning of states; finding productive engagements between these two sets of actors will demand imagination. In an age of greater individualization of politics, risk and uncertainty

(Beauregard, 2018; Beck et al., 2003), it is citizens that often emerge as those with an imagination born of 'the need to act' (Bhan, 2019: 13) in a world where club and state planners can appear paralysed.

Citizens might typically be thought to possess an imagination for the short term and the fine grain of the built fabric of cities. However, across global south cities, citizens have necessarily expanded into collective and longer-term actions for their respective neighbourhoods, their city and the global commons in light of the failures of clubs and states. If 'planning is a contested field of interacting activities by multiple actors' (Miraftab, 2009: 41), a purposeful urban planning mix might have significant potential at the intersection of state and citizen actors shown in figure 2.1. This potential has existed in the case of statutory planners working for more equitable outcomes in Cleveland in the US (Krumholz, 1982) and may yet be produced in the experiment with localism in the UK (see chapter 5). It exists in the emotional intelligence that has been little appreciated by academics and practising planners (Hoch, 2019) but which is vital to recognizing and empowering marginalized citizens.

The urban planning imagination of club actors has typically been visible at the middling scales of neighbourhoods, districts or self-contained settlements and in the middling time frames relating to the build-out of communities over several decades. Turning the undoubted resources and customer or special interest focus of planning by or for clubs to more consistently socially just, sustainable and inclusive ends remains a challenge and opportunity for the urban planning imagination. Club actors span the spectrum from for-profit developers of new communities with a keen appreciation of broad segments of consumer tastes to associations with an intense focus on and skill in advocating for minority interests, and we ignore either of these capabilities and imaginations at our peril. At the intersection between citizens and clubs are, for example, not only the socially minded *Baugruppen* housing developments (chapter 3) but also any number of less deliberative home-owner associations of gated communities.

Nation states have typically master planned at neighbourhood and city scales and offered strategic national spatial plans while simultaneously regulating with land-use and building codes at the finer grain most familiar to citizens. Nevertheless, many states across the global north and south now struggle with, or have retreated from, big-picture planning. Of the actors considered here, it is the state that appears least able to deal with the complex challenges of place making in the present. And yet 'it is in relation to the state that social change is articulated and enacted' (Roy, 2018: 145); it is precisely in those intersections between the urban planning imaginations of states, clubs and citizens shown in figure 2.1 that new urban imaginaries are to be found. The combinations of states and clubs in figure 2.1 find expression in the charrette method (chapter 5) and in the design competitions for publicly owned or acquired land in Scandinavian countries (chapter 4). Some of the cities that are most referenced at present – Dubai, Singapore – are powerful but limited amalgamations of club and state imaginations. Vacuums in statutory planning capacities in the global north and south are filled by the exceptionalism of master-planned club communities (Roy, 2005). The city is both a problem for and a prompt to government; questions of citizenship and the state come together in the city in ways which may yet see global south urban planning imaginaries increasingly take root across the global north (chapters 5 and 6), not least since the domestic and international diasporas that constitute some communities mean that citizens are ever more at the centre of networks, flows and virtual connections, through which the urban planning imagination can be mobilized with states and clubs in methods that resemble social learning (Bollens, 2002).

The overlaps in the imaginations of actors at the centre of figure 2.1 might considered a 'sweet spot' – one that promises the production of new urban planning knowledge called for in the 'southern critique' (Bhan, 2019; Lawhon et al., 2020) of the universal relevance of global north urban planning approaches. States and clubs have much to learn from citizens with respect to the appropriateness or frugality of urban planning interventions (chapter 7). The sharing and production of new knowledge among citizen, club and state planners can increase the options and negotiating skills open to the former (Anzorena et al., 1998) and generate pragmatic solutions to complex problems of ordering urban space, such as land sharing (Angel and Boonyabancha, 1988) and participatory budgeting, that have wider application (Carolini, 2015). At its best, the intersection of imaginations shown at the centre of figure 2.1 may see citizens, club and state actors share their respective expertise, unlearn some of what they thought they knew, and let go claims to exclusive substantive interest, wisdom and methods. However, we should guard against easy assumptions. At its worst, 'pick-and-mix' planning might amount to nothing more than an incoherent jumble of motives, principles and visions – an eclectic mix empty of any sense of the wisdom of urban planning. 'Spatial plans cannot and should not reconcile the multiple beliefs and expectations that come into play animating the places we inhabit' (Hoch, 2019: 2). Pick-and-mix planning may be rendered as a processual exercise that reconciles different actors' interests in lowest-common-denominator outcomes, as with planning for growth in England's 'Gatwick Diamond' sub-region (Valler and Phelps, 2018) or strategic spatial planning in Northern Ireland (Brand and Gaffikin, 2007).

The history and temporality of planning

'Modern town planning sprang from ... two different worlds, far removed from each other in time and space: the one embracing ideal cities and finite visible utopias, heavenly and earthly Jerusalems, perfectly formed works of art; the other composed of documents, manifestos, pamphlets and blueprints for new social orders' (Rose, 1984: 33). These twin aspects of urban planning can be seen in the comparisons of different planning systems presented in chapter 6: some are more abstract and ideal in their complete codification of rules; some define urban planning in more empirical, pragmatic and discretionary terms. Discussion of urban planning systems and cultures needs to move beyond history as interesting contextual background (Booth, 2011: 20).

Tilly (1984) distinguishes macro- and micro-historical levels of analysis where the former include urbanization, state making and bureaucratization. Little of the extant urban planning literature addresses itself to processes of macro-historical change; it instead speaks to the micro-historical level of encounters of individuals and groups or a meso-level of the institutional configurations of nations.[1] Yet 'national, international, regional, local and personal factors intermingle continuously' (Sutcliffe, 1981: 188–9) in the history of urban planning.

Macro-historical change: empires, economic systems and states

If, in the present, urban planning and its effects are hard to define (Reade, 1983; Wildavsky, 1973), in the longer sweep of history it is clear that 'no city, however arbitrary its form may appear to us, can be said to be unplanned' (Kostof, 1991: 52). An historical perspective can help uncover the ebb, flow and travel of planning ideas.

'Great cities with long histories are palimpsests, the developments of one era self-replenishing and half-replacing those of earlier times' (Corfield, 2013: 837). However, the term 'palimpsest' can evoke an excessive sense of continuity. There are discontinuities apparent in the making of cities – as with Mexico City's Plaza de las Tres Culturas (figure 2.2) – whether we are concerned with the spans of time taking us back to ancient history or those of an individual's lifetime. China may be exceptional as the one continuing urban civilization without permanent interruption (Morris, 1994: 1–2). Paradoxically, the continuity in Chinese civilization may be due to the ephemerality of its imperial city building since, unlike their Roman counterparts, Chinese cities were not built with monuments for eternity (Laurence, 2013). Evidence indicates the inert layering of development in the case of ancient civilizations where 'the past was not seen as something to understand on its own terms, but rather a source of ideas to reinforce contemporary ideas and practices' (Smith

Figure 2.2   Plaza de las Tres Culturas, Mexico City
Source: author

and Hein, 2018: 109). Modern empires, by contrast, drew on the past but also innovated.

The interactions of macro-historical layers reveal some of the breadth of the urban planning imagination. Different metaphors that have been associated with the city provide clues to the transformative potential of urban planning interventions. If the city is organized according to cosmological principles, it may be entirely resistant to urban planning. However, 'if the city is a machine that must function effectively, it is subject to obsolescence, and needs constant tuning and updating ... If the city is an organism ... it can become patho-logical, and interventions will be in the form of surgery' (Kostof, 1991: 16).

Capitalism, for all its seeming inevitability, is the product of an eventful history. The emergence of capitalism might itself be considered an event (Sewell, 2008: 530); an event by no means inevi-table or universal – an easily overlooked insight that must continue to inform alternative urban planning imaginaries, as I discuss in the next chapter. Nor does it totally erase preceding systems, as these continue to play into the urban planning imagination. For example, the UK liberal market in the exchange of land and property operates with a feudal pattern of land ownership and might be considered socialist in its allocation of the rights to development on land. Nevertheless, capitalism is a system that continues to 'colonize' not only activities (such as the production of culture) but also parts of the world (for example, societies across Africa based significantly on subsistence agriculture) that are distinctly non-capitalist. The tempor-ality of capitalism is 'composite and contradictory, simultaneously still and hyper-eventful' (Sewell, 2008: 517), producing both 'spatial fixes' and 'spatial switching' (Harvey, 1985) of investment in and out of cities.

Meso-level institutions of states

There is significant inertia in the built environment. The production of the built environment involves sunk costs in land and property which are imperfectly divisible or 'lumpy' commodities bought and sold in highly imperfect local markets. In addition, the development of cities is highly path dependent 'because of the complex of rules, configur-ations, and relationships of property/infrastructure/governance that are established in urbanization processes' (Sorensen, 2018: 42);

because, that is, of the differently instituted planning systems and cultures of nations that 'mediate competition over the use of land and property, to allocate rights of development, to regulate change and to promote preferred ... urban form' (ESPON, 2018: vii).

Sorensen (2018) develops a four-fold set of scenarios that highlight the evolution of the institutions that constitute national and sub-national planning systems and cultures. He identifies 'displacement', where the removal of existing rules and the creation of new ones are likely; 'layering' of new rules on top of existing ones; 'conversion' involving incremental change; and 'drift', or a failure to adapt policies to changed circumstances. Two of these processes – displacement and layering – have been apparent in the development of statutory planning systems and cultures.

The history of urban planning suggests that displacement has been something done across the global south as part of imperial expansion (Home, 2013), including the seizure of lands from indigenous populations and the imposition of norms of private property. In Australia (Jackson et al., 2017) and Canada (Blomley, 2014) these norms displaced ancient customary land-ownership relations and 'urban' planning as land management. Displacement has resulted from major political-ideological shifts such as the 'big bang' liberalization of land and property markets experienced in some East and Central European countries after the collapse of the Soviet bloc.

Statutory planning in many global north nations might be characterized by a layering of practices and responsibilities, as it has become a generally larger and better-resourced activity that has an expanding, complex range of responsibilities which require correspondingly elaborate divisions of labour. The simple world of the generalist urban planner, as he or she would have been trained at the end of the 1960s in the global north, has become more complex with the need to adjudicate on a range of complex technical evidence, specific new legislative and policy requirements and health and well-being aspirations.

Urban planning thus emerges as a matrix of institutions (Sorensen, 2018) underlining the centrality of urban planning to modern societies. Planning's institutions are beholden to economic structures and political forces and adapt chameleon-like to them. As such, 'planning and urban governance present an exceptionally dense and consequential set of institutions that is increasingly important for managing and regulating processes of urban change and capital investment in cities' (Sorensen, 2018: 42). The institutions of urban planning are a product of past actions (Salet, 2018) which provide a store of wisdom, and while the copying of prior responses has advantages (Hoch, 2019: 99), it may also prove a dead hand on the development of the urban planning imagination. When we view them in geohistorical context, we see both contrasts and commonalities in the evolution of these meso-institutions that are the statutory urban planning systems and cultures I discuss in chapters 6 and 7.

The micro-level

Modern urban planning in capitalist economies 'developed at some points in national leaps and bounds whereby local and historic practices were almost entirely irrelevant to its progress' (Sutcliffe, 1981: 207). One reason for this is the dynamism injected into local contexts by individuals. The effects of micro-historical processes of change should not be ignored when set against macro- and meso-historical forces. Those individuals involved in the planning of colonial outposts in Africa rarely stayed long – sometimes just a matter of weeks or months – but the effects of their visits were lasting (Home, 2015). Individuals were partly responsible for elements of cultural hybridity apparent by the 1700s (Bayly, 2000) and which continued to unfold in the history of modern urban planning under imperial powers (Nasr and Volait, 2003).

While international exchange of planning ideas and practices was intense by the early 1900s, Ward (2005) notes that, unlike today, the individual planning actors involved could hardly be considered part of a Global Intelligence Corps (Olds, 2002) or a Transnational Capitalist Class (TCC) (Sklair, 2001). The transatlantic foment in urban planning ideas and practices (Saunier, 2001) during the late 1800s and early 1900s was significant but hardly bureaucratized in the way such exchanges are today. Instead, the forces of exchange were aggregations of numerous independent study groups, exhibitions and conferences. Only a portion of these individuals were what Sutcliffe (1981) regarded as 'home-based' in outlook. The remainder were arbiters of increasingly cosmopolitan urban planning tastes.

Individuals remain influential in the shaping of cities and urban planning discourse. Individualism has been one key ingredient in some of the UK's most successful plans (Wray, 2016). In some cities and nations, both global north and south, it may be little exaggeration to suggest that political leaders have been the most influential citizen actors. Singapore's rise from third world to first can hardly be separated from Lee Kuan Yew's vision and influence (Lee, 2011). Mayors can be extremely important in the urban planning imagination and what it can achieve practically. New York City's Robert Moses provides an example of one person's mobilization of a municipal bureaucracy in the service of urban planning goals (Caro, 1974). Local urban planning agendas in a country like Indonesia are intimately related to the capacity and interests of mayors, to such an extent that rankings of local performance and policy emphases change markedly in a few years (Phelps et al., 2014).

There are charismatic architect–planners such as Santiago Calatrava who demand the absolute faith of citizens and politicians (Tarazona Vento, 2015). US architect-planners Andrés Duany and Elizabeth Plater-Zyberk have promulgated new urbanist formulations with the same sense of mission as Ebenezer Howard and Patrick Geddes. A global diaspora of expert consultants (Larner and Laurie, 2010; Prince, 2012) specializes in the substantive concerns discussed in the next chapter. These individuals operate within and draw sustenance from a now vast, globalized bureaucratic urban planning apparatus.

Citizens emerge as powerful vectors of transformation in the urban planning imagination in the case of neighbourhoods subject to major immigration flows and resultant ethnic concentration or 'superdiversity'. Here, planning impacts on the urban form can range from immediate needs for religious, ethnic or culturally specific facilities to the injection of broader, longer-term imaginations centred on place making and the remaking of nature (Gottlieb, 2007).

The temporalities of urban planning

Macro-, meso- and micro-level histories imply that multiple temporalities are present in the practice of urban planning. Indeed, it is this that demands the comparative look at urban planning: 'If townmaking, and urban life, are not a steady state of existence but surge and lapse in irregular cycles across the continents, alternative orders of human settlement should be given due attention' (Kostof, 1991: 31).

'Planning is ... a particular form of governmental technology through which social discipline, ritual, and rhythm are made present in social life, and in which time is materialized' (Abram, 2014: 129). The activities involved with urban planning often take far longer than the time it takes to build, which prompts Abram to question whether urban planning is an act that aids social adjustment to otherwise disruptive changes. Loss of information (Pollitt, 2000) or of wisdom and churn in urban planning policy (Rozee, 2014) may be products of the generalized speeding up of society (Bertman, 1998). Indeed, Friend and Hickling (2005) identify turbulence, urgency and overload as some of the very real pressures affecting urban planning processes. Interestingly, urban planning processes and plans themselves have become more, not less, various and abundant despite such pressures.

There are several different temporalities within any given planning system. There are the near instantaneous planning and urban service delivery decisions that are being made as a result of real-time data and visualization of it. But is this planning? Is it simply crisis management? Visualizations of real-time data convey nothing of urban planning's normative glance to a more just or sustainable future. There is a danger of being stuck in a permanent planning present of resolving day-to-day hassles rather than the envisioning of alternative urban futures.

There is the time frame on which plan making *ought* to take place if it is to command widespread support. There are separate political or bureaucratic cycles over which the preparing, making and adopting of planning documents *must* take place, and the time frames over which plans are monitored, evaluated and revised. It is clear that these different temporalities of planning themselves conflict, and this alone should alert us to the fundamental import-ance of the political and institutional support that urban planning needs to draw upon if it is to be anything at all.

There are the spans of time over which changes happen outside the planning process itself. Forecasts of population growth, housing needs and residential land-use allocations can often be exposed by fluctuations of business and property investment cycles, let alone Kondratieff long waves which generate severe economic crises.[2] Here the influence of developers as club actors on and in urban planning processes orchestrated by the state and affecting citizens becomes apparent (Charney, 2012). The etymology of the word 'crisis' in the Greek *krisis* – meaning decision – is itself suggestive of planning. Crises are shared experiences calling forth the collective actions that we take as the signature of urban planning. It may be true to say that 'crises have become a normative expectation in contemporary societies' (Hall and Ince, 2018: 9). Certainly 'disruption' now figures prominently in writing on the implications of big and real-time data for urban planning (Batty, 2018; Hall and Tewdwr-Jones, 2020).

To be effective, planners need to be able to contribute on a given issue for a period of time (Krumholz, 1982). The problem is that churn is a feature of the careers of trained urban planners regardless of which citizen, club or state actors they work in the service of. Even the lesser time frames discussed above may exceed the tenure of individual planners. The meso-historical time frames needed to resolve some urban planning challenges far exceed those of particular institutional configurations of statutory planning, let alone the lives of individual citizens or trained urban planners. For example, 'even though the science and use of scenarios in climate change projections might suggest the need for a long-term view, UK planners and planning authorities have ... been inhibited in taking a long-term perspective or in engaging with futures thinking' (Wilson, 2009: 223). In macro-historical terms, the urban planning imagination is revealed as seeking for resilience and gradual adjustment to climate change and events, but it is questionable whether and how it can respond in the short term to the increasing frequency and greater severity of climate events (Halsnæs and Laursen, 2009: 83).

Across the global south, time is a resource that citizens know how to exploit. Inaction by states, or their inabilities to fully enforce regulations, become opportunities for income generation in tactical adjustments by street vendors (Recio, 2021). Waiting for the state to act opens the way to the autoconstruction of housing (Oldfield and Greyling, 2015). In both instances the citizen–state relationship is reconstituted. The temporal interstices of urban planning are essential to understanding some of the creative energies of global south citizens in resisting some of the institutionalized divisiveness of statutory planning.

Statutory planners are adept at making adjustments across increments of time in trying to solve the complexities of urban development (Hoch, 2019: 51) and have become more adept at utilizing the interstices of planning processes when sanctioning temporary uses of spaces (Madanipour, 2017). Across the global north, temporary uses of urban space can introduce experiences that rub up against the fixtures of statutory urban planning practices and in which difference is affirmed (chapter 4).

The geography of the urban planning imagination

For much of its history, urban planning has been conceived and practised as an activity operating on the physical appearance of the city – its land uses and building forms – as a set of bounded places. This is the city as container of people and activities at whatever scale we choose to analyse it – neighbourhood, metropolis or megalopolitan region. It is understandable even from today's vantage point why those studying and planning cities would think of them as containers – ancient and medieval cities were defined by their walls and gates. Over time a relational, geographical sense of urban planning of and for cities has come to the fore in several metaphors (figure 2.3) that provide the basis for new urban planning knowledge and practice, recognizing any number of social, economic and environmental processes operative on and through the city.

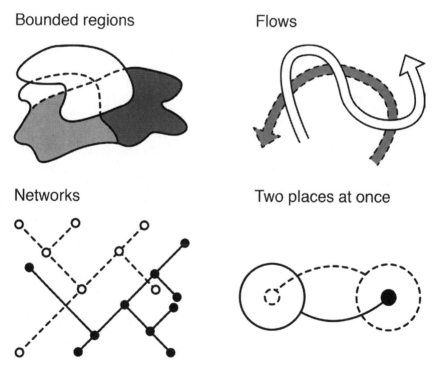

Figure 2.3 Four geographical metaphors
Source: Thrift and Olds (1996)

Bounded places: neighbourhood, city, region, nation, metropolis and megalopolis

Natural regions are easily defined in terms of mountain chains, river basins or climatic zones, but places constituted partly or wholly on the basis of human activities prove less stable and evolve in ways that present challenges for the urban planning imagination.

The scalar sensibility inherent in urban planning thought and practice remains important to discerning differences in the style of urban planning within nations. As we ascend scales, urban planning typically becomes less concerned with regulation of the use of space and more strategic, framework-setting and visionary (ESPON, 2018). A scalar sensibility is important to appreciating some of the international contrasts to be found in urban planning. Thus, 'Perhaps the major difference between plans and planning frameworks in the United States and Europe is the issue of scale' (Knaap et al., 2015b: 511). Where issues of land-use regulation and urban design at the neighbourhood scale exercise minds in the US, in Europe they are taken for granted given the greater historical preservation offered by regulatory frameworks.

Sites and neighbourhoods

Site planning might be understood as something circumscribed to scales anywhere between an individual building and a neighbourhood of a city and is especially relevant to architects and urban designers. The site scale also figures in urban planning's contributions to sustainable development, since regulatory urban planning has found itself in advance of the development sector when driving revisions to building codes (Rydin, 2009). An example here is the 'Merton Rule' (named after the London Borough of Merton): it instituted a 10 per cent on-site renewable energy policy for all new development and was subsequently taken up by 170 local authorities. The adoption and enforcement of environmental certification systems such as BREAAM and LEED have been 'normalized' in the property and site selections of major businesses as a result of statutory urban planning.[3] Urban planning's potential to address environmental concerns is therefore often registered in the eco-credentials of celebrated constructions, such as the BedZED development in London.

The neighbourhood scale has retained an underlying appeal within urban planning practice despite the ambiguities of the idea of 'community' with which it is associated. For Clarence Perry, who elaborated the neighbourhood concept associated with the Radburn design by Clarence Stein and Henry Wright, a neighbourhood possessed 'a certain unity which is quite independent of political boundaries' and could be 'regarded both as a unit of a larger whole and as a distinct entity in itself' (Perry, 2011 [1929]: 488), and perhaps the key element and innovation in this particular site plan was the hierarchical system of roads in which boundary roads took bypass traffic. In the US, Garden City ideals found their way into the Radburn neighbourhood design (Talen, 2005). However, in the curious manner that urban planning ideas have travelled, 'the Radburn exemplar, not very influential in its country of origin ... would become instead a standard point of reference in Sweden and in Britain's New Town programme' (Kostof, 1991: 82).

The neighbourhood scale is prominent in new-urbanist-inspired settlements and town extensions. In the US, the Seaside and Kentlands developments have been central to the success of the new urbanist movement (Passell, 2013). In the UK, Leon Krier's Poundbury has achieved notoriety as a planning model. With its mixed uses, mixed plot sizes and tenures, Poundbury is one carefully curated example of the desire to fashion neighbourhood-level community: 'The development approach attempts to simulate the simplicity of design yet diversity associated with the gradual organic development seen in historic Dorset towns and villages, albeit within a greatly compressed timeframe of 10–15 years' (Thompson-Fawcett, 1998: 182). The attempt to reproduce existing settlement form has seen the criticism of inauthenticity levelled at it. However, it has travelled as a planning model of the sort I discuss in chapter 7, 'being transformed in morsels to numerous other projects ... visits to Poundbury are continual, both by British groups and those from abroad' (Thompson-Fawcett, 1998: 185).

Rural associations have been folded into the real and imagined experience of urban *kampung* (village) life in Indonesia as pre-existing rural communities have been absorbed into, or developed as new neighbourhoods of, rapidly expanding cities. However, these essential neighbourhood reference points are incredibly diverse. In the city of Bandung, Benjamin et al. (1985) found that urban *kampungs* varied in size from under 200 families (perhaps 1,000 people) to over 60,000 families (perhaps 300,000 people). Moreover, the social capital of the *kampung* is ambiguous in its meaning for development (Woolcock, 1998) and the urban planning imagination.

Machizukuri, a concept of neighbourhood planning seemingly different from the top-down civil engineering tradition, has emerged in Japan. Its origins reflect a conjunction of trends – such as the rise of grassroots movements alongside governmental decentralization – and the imperative to recover and reconstruct after the Hanshin earthquake of 1995. Ambiguities surrounding the concept itself mean that both authentic and less authentic examples can exist in the same city, as in the cases of the Mano and the Rokkō-michi Station South Areas in Kobe. Thus, Mano – one of the best examples of *machizukuri* practice in Japan – was by 2002 'still a long way from the ambitious "future image" proposed in 1980' (Evans, 2002: 452). This particular neighbourhood has become emblematic of the community planning process in Japan as a whole, as 'There seems to be a feeling that, if Mano fails, community planning in Japan will be that much poorer' (Evans, 2002: 456).

The neighbourhood scale of urban planning was important in socialist cities. In the former USSR, the microrayon (micro-region) formed a key neighbourhood-scale building block of the Soviet city (French, 1995). The vast majority of the extant urban fabric of Chinese cities, as they stood as recently as 1980, had been built in a patchwork of walled and gated work units or *danweis*. As neighbourhoods of the socialist city in China, *danweis* were self-contained, being places of work and residence with all the necessary urban services – schools, clinics, recreational facilities, shops and canteens. Moreover, and in contrast to the aspirational ideals of the gated neighbourhoods of liberal market economies, they were real communities given the immensely strong identification of people with their *danwei*.

Historic reference points for considering neighbourhood-scale planning have been joined recently by a model pioneered in one of the more densely developed cities of East Asia. Singapore's Housing Development Board pioneered and exported its neighbourhood planning concept to China and has continued to refine it, to cater for higher densities of dwellings and population, not least by drawing on the experience of its export to and implementation in China (Miao, 2018b).

Regional planning

One of the earliest visions for regional planning described how it brought together several fields of expertise – economists, surveyors, town planners – in a 'product of a "composite mind"' and distinguished regional planning from metropolitan planning, which 'assumes a continuously expanding metropolis as an inevitable if not desirable condition' (MacKaye, 1962 [1928]: 35, 39).

However, the problem of defining an appropriate scale for urban planning to operate at is at its most apparent here. With the exception of a few countries, regions, as ambiguous scales ranging from the metropolitan to below the national, have frequently failed to capture the popular or political imagination. Perhaps as a result, and in contrast to MacKaye's hopes, regional planning has often been reserved for particular, not composite, purposes. These types of regional planning include physical and economic planning, allocative planning and innovative or development planning, multi- or single-objective planning, or indicative or statutory planning (Glasson, 1978). That is, 'regions are an intermediary level, both territorially and functionally, and their power depends on their ability to integrate various levels of action, on their knowledge and mastery of decision making networks' (Keating, 1997: 393). Rarely, it would seem, have regional institutions been able to mobilize power, let alone consistently over time.

In the UK, weak reforms were made in the 1990s to regions established much earlier to administer policies and disburse funds, while regional spatial strategies that were a decade in the making at the start of the twenty-first century were abandoned overnight by central government, indicating just how insecure regional planning can be. In the US, the Regional Planning Association most clearly espoused the logic of regional-scale planning and was successful in promoting the establishment of the Tennessee Valley Authority (TVA). The passing of the TVA Act in 1933 granted the TVA statutory responsibilities pertaining to water resource management and an ambiguous remit of regional development and planning. In the course of its practical affairs, the TVA was soon shorn of this latter remit (Schaffer, 1986).

Physical geography ensures that some instances of regional planning persist and have effects. Only 16 of the 229 different types of plan in existence found in one recent survey (OECD, 2017a) did not fit into the scalar hierarchy of national, regional and local. These were, for example, Turkey's coastal plan, the Czech water catchment plan, and the German open pit lignite mining plan. However, functional urban regions, while they can be defined on the basis of, for instance, labour market catchments or commuting patterns, have rarely gained traction and permanency in urban planning in themselves beyond being references for the collection and reporting of data. The sorry tale of the failure of proposals to reconstitute archaic local government boundaries on travel-to-work areas under the Redcliffe-Maud Commission (reported in 1969) – despite a history of frequent local government reorganization in the UK – is testament to that. The rejected proposals have been looked back on fondly as something of a missed opportunity (Leach, 1997).

While something of the sentiments of regional planning has been apparent with respect to the metropolitan areas that planners and politicians accept as coherent functional regions and that most citizens can identify with, local politicians have been reluctant to change constituency boundaries, while national politicians have always hotly debated the electoral implications of bigger units of urban government. Proposed local government reorganizations – such as those in the 1960s and 1990s in the UK – founder on these issues (Tewdwr-Jones and Allmendinger, 2006). In the US, the theoretical idealism of regionalism has given way to a more pragmatic and limited sense of metropolitan regionalism. This is both more geographically pragmatic in focusing on major urban centres and their hinterlands (of 50,000 population or more) and more substantively limited – being driven by federal funds directed towards transportation planning in approximately 400 Metropolitan Planning Organizations across the US (Salkin, 2015).

Megalopolitan realities

Cities did not emerge in splendid isolation but from the start developed within systems (Smith, 2019), so that city networks of megalopolitan scale and organization are not new. Archaeological research suggests that the central lowlands of Guatemala may have been home to between 7 million and 11 million people over 1,200 years ago in a system of settlements extending across an area of 95,000 km^2 (Canuto et al., 2018). Nevertheless, the term 'megalopolis' emerged in the early 1900s to decry the increasing scale at which urbanization was taking place (Baigent, 2004) and resurfaced later to describe the reality of large-scale urbanization along the eastern seaboard of the US (Gottmann, 1961). Gottmann later defined a megalopolis as a functionally interdependent system of cities of at least 25 million population acting as both an incubator of new economic activities and a national-international hinge of trade (Li and Phelps, 2018).

Today, many national economies show signs of organization at the megalopolitan scale. The world's forty largest mega-regions account for two thirds of economic output and 85 per cent of innovation (Florida et al., 2008). Megalopolis is a concept with renewed salience as a descriptor of the contemporary scale of urbanization and economic functioning of the US (Nelson and Lang, 2011) and of China, where one can travel by road or rail from Shanghai westwards through a near-continuous urban landscape to Suzhou, Wuxi and beyond. And yet megalopolis has limited appeal as a scale for urban planning. The idea found some favour in Japan in the 1970s (Hanes, 1993). Today, active megalopolitan-scale planning efforts in China are apparent in the Pearl and Yangtze River Deltas, though their purchase on patterns of urbanization and infrastructure development remains unclear in a context of rivalry and duplication of functions among cities (Wu, 2015). In Europe, the Randstad area of the Netherlands is a recognizable and coherently planned constellation of cities but lacks the scale to be considered megalopolitan.

Less clear still is whether the morphological and economic appearances of 'ecumenopolis' (Doxiadis, 1962) – urbanization stretching across continents – will ever capture the political or planning, let alone popular, imagination. Something of this ecumenical scale of economic ties and cultural connections is made explicit in China's 'Belt and Road Initiative' (BRI).[4] This scale is certainly apparent in the narrower joint planning efforts of states and multinational enterprise clubs interested in ensuring the smooth functioning of today's logistics corridors that unprecedented levels of international economic integration rely on (Cowen, 2014).

National planning

For all the valid and mounting criticisms of 'seeing like a state' associated with statutory urban planning, we remain in a world where nation states and senses of nationhood continue to form, notably as a product of the 'rescaling of the nation state' by way of governmental decentralization and devolution (Brenner, 1998). And yet, ironically, given statutory urban planning's intimate relation to projects of nation-state building (Yiftachel, 1998), the national scale of urban planning has only rarely been of importance. Ten of thirty-two of the world's wealthiest nations included in a recent OECD survey prepare neither general spatial or land-use plans nor guidelines on land use (OECD, 2017a: 15). National spatial planning has never been a strong feature of liberal market nations such as the UK and the US. The federal government of the US exerts little direct influence in matters of urban planning and a national Land Use Planning Act drafted in 1970 was never adopted (Salkin, 2015).

Even in Europe, the birthplace of the modern nation state, national-level urban planning has been strong in only a few countries and then only briefly. Discussions of the distribution of population and employment in the UK initiated in the 1920s informed a national political consensus in pre- and post-war urban planning (Hall and Tewdwr-Jones, 2020). However, when a national spatial plan was forthcoming in the 1960s it was brandished briefly before being quietly put away in a drawer somewhere in Whitehall. In continental Europe, the hierarchical plan ideal is strong but is rarely achieved in practice, and national plans have tended to become less important. National spatial planning now exists more as a taken-for-granted or implicit context shaper rather than an explicit frame of reference. In the Netherlands, in the Randstad and its counterpart green heart, 'planners found a coherent mental map of their country and its development' (Faludi, 2015: 273) that now hardly needs explicit representation in any national plan, although even here there are proposals to move away from that established planning form towards more fuzzy or softer forms of planning (Balz and Zonneveld, 2018). Across Europe only a few countries have increased planning powers at the national level in the face of demands for decentralization and devolution (ESPON, 2018: viii). Across the EU, the nature of national planning has changed. It has 'moved increasingly away from spatial, comprehensive, and distributive roles towards sectoral goals, strategic national interests, economic competitiveness, and more recently, dealing with climate change' (Knaap et al., 2015b: 505). Worryingly, national planning has moved away from long-term concerns requiring an integrative and synoptic perspective towards particular short-term preoccupations.

Networks, flows and virtual urban planning

Key tools at the urban planner's disposal have been land-use or zoning plans and building codes. Such 'codes' seek to order the firmly bounded spaces or geographic scales described above. Where once land-use plans might be considered to have a beauty and seductive power of their own, they may now present a limited and increasingly unappealing visual medium, since:

> planners and plans have been criticized not merely for trying to 'order' the dynamic and inherently disorderly development of cities and regions. The concepts that have been used ... are seen to reflect a view of geography which assumes ... contiguous space ... that physical proximity is a primary social ordering principle and that place qualities exist objectively, to be ... made by physical development and management projects. (Healey, 2004: 47)

The reliance on land-use plans to shape physical developments was a peculiarly twentieth-century phenomenon. Set within the fixed administrative boundaries of individual local governments, each plan – where it existed – was prepared over years of data collection, analysis and projection regarding demographic shifts, housing requirements, employment trends, transport use, commuting, health provision, educational attainment, retail spending, leisure patterns, tourism and the like prior to public consultation, by which time the plan was likely to be out of date.

In a world in which most development is now undertaken by citizens and clubs it may seem anachronistic that we still rely on plans produced by the state. Moreover, many of the trends affecting cities today – the 'gig' or digital 'platform' economies – have only indirect relationships to land use. Equally, some of the most influential and seductive of 'plans' have not been land-use or zoning plans, though they have still portrayed fixed arrangements between morphological elements. It was the spatial form of Ebenezer Howard's Garden City idea that became the ideal, not its land-ownership and governmental formulations. What will be the fate of these sorts of plans?

In an era when data can be produced on any number of urban trends in seconds and made publicly available quickly rather than in months and years, is the traditional land-use plan now spent? The rigidity of the spatial imagination associated with such plans sits uncomfortably with urban planning being a future-oriented activity attuned to aspirations and, as such, subject to continual revision. And yet, there are paradoxes here. For plans have continued to proliferate, albeit in new and varied styles, and the appetite to be modern among many nations and peoples (Ferguson, 2006) will ensure that land-use plans endure.

While the urban planning imaginaries associated with bounded regions are well established, the planning imagination associated with the networks, flows and virtuality (or synchronicity) of contemporary life is immature by comparison. A problem for urban planning – given its dual analytical and normative aspects – is whether these new imaginaries are capable of being rendered visually and rhetorically in ways that are seductive or practical as guides for intervention. Can those planning theories that seek to mobilize the communicative, collaborative or deliberative (Forester, 1993; Healey, 1997; Innes and Booher, 2010) properties of networks, flows and virtual connections of citizen, club and state actors ever appeal or be harnessed to practices of planning actors? I discuss these properties further in chapter 5.

Cities as nodes within networks

The contemporary urban planning imagination may recognize cities as less enmeshed in a hierarchical set of bounded places orchestrated by nation states than as present within the horizontal networks of connections between a vast number of cities and associated citizens, clubs or states. Plans are capable of expressing the character of the city as a network of nodes. This much is familiar in the topological maps of the metro systems of London, New York or Paris. These maps are also pieces of art though many of them can be illegible to all but the initiated.

This perspective interprets flows of money, commodities and people between cities as accumulated stocks – including investments made in the name of urban planning – which in turn promote inertia in a hierarchical network of world cities (Taylor et al., 2002: 2388). The key attributes of this system were outlined by Friedmann and Wolff (1982): (1) world cities articulate regional, national and international economies into a global economy and serve as the organizing nodes of the global economy; (2) elsewhere, major regions of the world are excluded from these cores and live instead in an economic periphery or semi-periphery; (3) world cities are large urbanized spaces of intense social and economic interaction; (4) the group of world cities is itself a hierarchical system of major cities – and they can be ranked; and (5) the elite of world cities increasingly constitute a transnational social class – a social class that is the product of global capitalism and that acts to ensure its survival.

Notwithstanding some of the silences of world-city analysis, which I return to in chapter 7, it is useful in drawing attention to questions of social polarization (Friedmann and Wolff, 1982: 322). Saskia Sassen (1991) later developed this line of thought, arguing that structural transformation towards service industries and the crowding-out of manufacturing in many global north world cities has seen the loss of skilled manual occupations (middle-income). These generalities have specific ramifications for urban planning and planners in world cities, where housing affordability and investment in schools, parks, transit, hospitals and the like are vital to reproducing the success of those same city economies.

The world-cities literature addresses itself almost exclusively to the economic rather than the political or planning networks of cities. Nevertheless, if 'city powers ... are mobilized through networks' (Allen, 2010: 2898), then, by extension, the urban planning of cities might be said to be networked. City networks appear to have proliferated in number, diversity and depth, with implications for urban planning thought and practice that have barely been considered (Davidson et al., 2019). Despite the possibilities, there are obstacles to information pooling let alone collective action across networks of cities (Miao and Maclennan, 2019). The statutory planning efforts of local governments across the world rely on evidence bases that are now often produced by club actors. How, then, will policy experimentation in networks of cities shape the production and availability of alternative planning knowledge?

Networks of cities now encompass large parts of the urban population and mobilize significant resources. One quarter of global gross domestic product and 650 million people are accounted for by the C40 Cities Climate Leadership Group, for example, while US$100 million of funding was associated with the Rockefeller Foundation's 100 Resilient Cities programme (Davidson et al., 2019). As diplomatic efforts (Lauermann, 2018), city networks have been an important means of lobbying national and supranational organizations for funding and promoting individual cities and their policy experience and expertise. The 'Barcelona model' emerged from the city's celebrated regeneration efforts in hosting the 1992 Olympic Games, to such an extent that the city of Barcelona now participates in numerous city networks. City networks have been seen to have been important to environmental policy experimentation (Bulkely and Castan Broto, 2013), as I discuss in chapters 3 and 7. The UN's Sustainable Development Goals have helped replace perceptions of the city as a source and container of defined 'urban problems' with perceptions of the city 'as a hub, driver and node through which all sorts of global challenges can be addressed in practicable ways' (Barnett and Parnell, 2018: 33).

Flows: the metabolism of cities

Urban planning has long wrestled with the relationship of the settlement of land to the flows under, over and across it that affect the health of citizens and that are needed to sustain settlement. The need for the healthy metabolism of cities that informed early modern planning drew inspiration from already mature water, waste and metro-system engineering achievements. However, flows arguably take on new individuated meaning in an information-rich age (Castells, 2005) with implications for the balance of citizen, club and state planning actors as they continue to shape our cities.

The supply of water and the removal of waste have been central to ancient and modern urban planning alike. They were the early and universal concerns of modern urban planning as it developed from the 1800s onwards in the global north, as I discuss in the next chapter. Cities such as London and Paris pioneered the introduction of city-wide water-supply and sewerage systems, and these were rapidly exported to overseas colonies as perhaps the ingredients absolutely central to the planning of new and expanded cities.

The same concern with planning for flows in and through the city emerged in connection with the desire to improve the efficiency of movement of goods and people. The first mass public transit system in the world, London's Metropolitan Railway, opened as early as 1863. Later, in the 1950s and 1960s, American cities were to pioneer planning to increase efficient movement within cities, catering on a massive scale for car use. The European Commission has sought explicitly, with the funding of transport networks, to fashion a Europe of flows (Jensen and Richardson, 2004), though policy and discourse for this continental space of flows continue to rub against discourse, practice and politics at city, regional and national scales (García Mejuto, 2017).

Diasporas and international trade were the two main vectors in the ancient world (Clark, 2013: 5). They re-emerged in a first global economy of the late 1800s to early 1900s and in a second global economy of the 1980s to the present (Jones, 2005). In particular, the urban planning imagination in many cities is being informed by intensified flows of people – superdiverse citizen actors in some cities. Urban planning systems are faced with the bifurcation of these flows of citizen actors into mass domestic and transnational flows of low-skilled, low-wage labour and flows of labour market and consumer elites that form part of a TCC (Sklair, 2001) that I discuss in chapter 7.

China has a long history of internal migration that continues into the present in altogether more massive flows from exporting rural provinces. Migrants arrive as floating workforces without full rights as citizens in the booming cities of the eastern seaboard. An international labour market and offshore manufacturing hinterland have been planned for Singapore since the 1980s. The trade and labour market arrangements agreed under the Indonesia–Malaysia–Singapore Growth Triangle (IMS-GT) significantly underpin the continued growth of an otherwise land- and population-limited city-state economy. Each day, thousands of workers commute into Singapore from Indonesia and Malaysia, while flows of executives and skilled workers go in the other direction to factories in Johor and Riau. The IMS-GT simply exceeds other transnational metropolitan or megalopolitan economies such as Tijuana–San Diego–Los Angeles in terms of the planning involved.

A diverse, internationally mobile elite have effects on urban planning systems and cultures in world cities. They include both those who speculate in land and property and well-paid workers deployed by multinational enterprises. These flows can blinker the urban planning imagination. Local urban planning systems have licensed processes of gentrification, resulting in the segregation of elite global populations into favoured quarters within world cities (Butler and Lees, 2006). Away from cities such as London and New York, the same condoning of inequality in urban planning systems is registered in the permitting of gated residential communities for global elites in Johannesburg or Guadalajara that promise the 'splintering' of urban space (Graham and Marvin, 2000). The impacts can be triply specific in that they focus local urban planning attentions on the redevelopment of particular city spaces, often in the form of mixed-use mega-projects, in many instances with a preference for iconic architecture. Mega-projects have been controversial for their loading of financial risk upon the public sector (Flyvbjerg and Sunstein, 2016; Tarazona Vento, 2017) and the way in which 'special-purpose delivery vehicles' have been used to circumvent local planning procedures and reposition cities for moneyed outsiders (Swyngedouw et al, 2002: 545–6). Many world cities are currently being remade not in the image of relatively inclusive paternalistic states or large numbers of citizens but in a troubling state–club fusion catering to the wealthy.

A heightened sense of inter-urban competition is signalled in these and other flows and has prompted plans which have become more indicative than previously; sharp lines and clear land-use allocations replaced with fuzzy boundaries and sweeps of likely development. The visual imagery now implies dynamism often beyond the jurisdiction of the plan itself. This indicative turn is associated with changes in the temporality of plans and, of course, the abandonment of any sense of their binding nature. Something of this is seen in the key visuals associated with recent metropolitan plans and their revisions (see, for example, Helsinki 2050, Canberra 2030, New South Wales 2050). It is named in what Allmendinger and Haughton (2009) refer to as the 'soft planning spaces' that have emerged at the subnational scale in the UK, where those spaces involve deliberate attempts to introduce new and innovative cross-sectoral and interterritorial ways of thinking in urban planning.

Virtuality and synchronicity

While similar or identical tendencies in societal constructs are commonly taken as evidence of a world society (Meyer et al., 1997), these same tendencies have been apparent historically. Thus, 'we need to ask if ... similarities ... are the result of common but autarkic human responses to the structural processes of urbanization ... and how far they are the effect of "connectivity" – the transfer of cultural, commercial and other ideas' (Clark, 2013: 4).

There is historic evidence of synchronous but independent developments in urban systems in different parts of the world (Clark, 2013; Kostof, 1991). The grid is one geometric organizing principle used in the planning of cities. 'Far more than a representation or signifier, the grid is a *world-making device* that literally brings new worlds into being' (Rose-Redwood and Bigon, 2018: 3, original emphasis), as seen in the case of Melbourne in chapter 7. While the form may be universal, variations are apparent. The rectilinear grids of the ancient Roman and Chinese empires were physically similar but produced from contrasting creative mentalities (Laurence, 2013). The grid provides evidence of synchronicity but does not submit to simple stories of origin or imposition (Rose-Redwood and Bigon, 2018).

Today, the similarities that we notice between cities are more likely to be a product of the virtual connections between citizens, clubs or states in webs of exchange of images, ideas, practices and methods facilitated by information and communication technologies (ICTs). Built-environment professionals – architects, realtors, planners – have always borrowed but are now able to scan the world far more rapidly than in the past. Will planning be little more than 'cut and paste' or 'photoshop planning' (Rapoport, 2015)? Global perceptions of a city such as Dubai – one of the 'born global' cities and planning systems discussed in chapter 7 – are derived from very particular representations of it (Elsheshtawy, 2009). Indeed, the prominence of Dubai in popular perceptions and rankings has been consciously planned through the construction and projection of the symbolic power of high-rise architecture (Acuto, 2010).

The implications of ICTs for the built form are all the harder to predict and plan for because rates of take-up and penetration of successive technologies providing for physical and virtual mobility have accelerated. It took sixty-eight years for half of the US population to make use of telephones, seventeen years for them to have a car and just seven years – the time it might take to prepare a local plan – to be using the internet (Kellerman, 2012: 77). The same contradictions between the enormous fixity of infrastructures that provide for mobility remain with the widespread deployment of ICTs. The danger is that 'the infrastructure of mobility, and the production and function of built forms could become the taken for granted bits, jettisoned from consciousness' and from judicious planning, in contrast to a 'digitally constructed, rationally planned set of activities' produced and consumed by innumerable individuals (Buliung, 2011: 1377).

The increasingly vicarious character of the contemporary urban experience promises unthinking imitation in some mixes of the urban planning imagination, and that is a concern for all it would entail about the production of good or better places.

Conclusion

The urban planning imagination is found among actors with different appreciations of, and attachments to, time and place. This presents a challenge and an opportunity, given the vast international reserve of urban planning imagination that might be tapped.

It appeared to Cherry (1996: 226) that 'the difficulty for town planning today is that intellectually it has ground to a halt. Its qualities of imagination for things that might be, and the internal energy to achieve these things, are squeezed out of a system where other matters get in the way' (Cherry, 1996: 226). Proliferation of plans may simply reflect the further bureaucratization of statutory urban planning and the individualization of politics in a second modernity (Beck et al., 2003), with different citizen, club and state actors talking past one another. Yet the urban planning imagination continues to flourish in a proliferation of non-traditional plans, in projects and in collaborations among citizens, clubs and states, and it will not be contained institutionally or professionally in the way it was in the twentieth century. All of this provides grounds for optimism surrounding the future of urban planning. The mixed character of urban planning has been apparent over the millennia, but if it is to be more than the lowest common denominator in the present age, then it is vital that it mobilizes sensibilities that seek awareness, complexity and variety in the shaping of places.

3 Substance: what are the objects of planning?

Introduction

'The fundamental challenges of building and sustaining human settlements have not changed significantly for centuries. Humans need shelter, sanitation, transportation, nutrition, social interaction, and economic production. The relative urgency of these challenges, however, has changed over time' (Knaap et al., 2015a: 1). Indeed, 'there is an interdependence of people in the city for the most fundamental human needs of water and food' (Smith, 2019: 12), though concerns have rightly been voiced over universal urban planning approaches to providing for them (Watson, 2009). These concerns coexist with an acknowledgement that in the global north, 'The emphasis in planning systems and strategies has shifted progressively from managing urban extension, remodelling the 19th Century industrial city, and servicing territories with adequate infrastructure systems, to a concern with environmental conservation, urban quality and development of the economic assets of places' (Healey and Williams, 1993: 702). Historical perspective allows us to see how changes in the substantive concerns of planning are layered over its apparent continuities for these enduring needs. The hundreds of specializations taught across North American planning schools mean that identifying the substantive footprint or boundaries of urban planning is challenging (Sanchez and Afzalan, 2017: 71). Some of the substantive concerns of urban planning as it is taught in accredited university programmes in North America have been discussed by Brinkley and Hock (2018) and are shown in figure 3.1.

I discuss most of these substantive concerns of urban planning and the delivery of societal *outcomes*. Ironically, these concerns are ones that modern urban planning has not always been well-equipped to

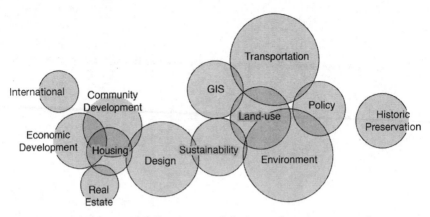

Figure 3.1 Thirteen specializations and their overlaps in twenty-six accredited US planning schools
Source: Brinkley and Hoch (2018)

deal with. A recent survey laments, 'it should be ensured that public policies outside of the domain of spatial and land-use planning do not run against land-use and related objectives' (OECD, 2017b: 18). Despite its synoptic potential, then, the urban planning imagination struggles to mediate between these substantive concerns. Acting in isolation, citizens, clubs and states have each tended to focus on and be sources of expertise regarding particular substantive concerns. The search for productive new mixes of the thought and practice of different actors demands that international perspectives will need to move to the centre of figure 3.1.

Shelter

Shelter is a basic need recognized by the UN through its Habitat II Agenda and its Sustainable Development Goals in a commitment to the 'full and progressive realization of the right to adequate housing' (United Nations, 1996: para 61, p. 34). Shelter is the issue more than any other that reveals the citizen agency involved in urban planning – an agency that sits sometimes uncomfortably with the agency exercised by clubs and by states.

Cities without slums: seeing like a state

In 2003, UN-Habitat estimated that 32 per cent of the world's urban population were living in slums. The picture painted is one of near-incomprehensible scale:

> Hundreds of millions of urban poor in the developing and transitional world have few options but to live in squalid, unsafe environments where they face multiple threats to their health and security. Slums and squatter settlements lack the most basic infrastructure and services. Their populations are marginalized and largely disenfranchised. They are exposed to disease, crime and vulnerable to natural disasters. (World Bank, 2013: 1)

Against this backdrop the UN launched its 'cities without slums' initiative in 1999. The challenge to conventional global north urban planning thought and practice presented by this scale of informality is immense, since 'the attractiveness of these kinds of locations for poor households is that they can avoid the costs associated with formal and regulated systems of urban land and service delivery' (Watson, 2009: 2265).

Well intentioned, the invocation of the term 'slum' is problematic, since 'If a slum is a relative concept, viewed differently according to social class, culture and ideology, it cannot be defined safely in any universally acceptable way. Nor is the concept stable across time because what we consider to be a "slum" changes' (Gilbert, 2007: 700). The poor physical appearance of housing and land use says little about the people inhabiting that housing. Numerous slum-clearance projects testify to the social and economic problems left in their wake. The comprehensive redevelopment of UK cities after 1945 left people with better housing but in localities physically remote and lacking in facilities. Comprehensive redevelopment plans were drawn up for 345 British towns and cities under the post-war new town and country planning system by 1962, with 100,000 people per annum being moved at the height of slum clearance during the 1950s (Robson, 1988). The numbers are staggering enough, but they pale in comparison to the numbers that future slum clearance across the global south might seek to dislocate.

Across the global south, 'as slums exist currently, they are teeming with life, social networks and economic linkages. It is often impossible to re-create these livelihood options and possibilities outside of highly fluid and malleable physical conditions that are best afforded by informal areas' (Pieterse, 2008: 57). Informality can be regarded as resourcefulness and social networks of action (Roy, 2005). Indeed, some slums have become famous. Dharavi in Mumbai is not only a residential agglomeration of over 80,000 structures and 700,000 people covering 2.5 km² but also a concentration of industries with an estimated turnover of $US665 million (Dyson, 2012). 'Slum tourism' exists here and in the favelas in Brazil. Peruvian economist Hernando De Soto Polar's idea – of formalizing the property rights associated with informal settlements in order to increase the flow of credit to the poor – recognized this resourcefulness, but the evidence from Peru suggests that formalization of property rights enabled only 1.6 per cent of households to leverage finance (Pieterse, 2008: 49).

Urban informality highlights the issues of compensation and betterment I discuss in the next chapter. In Jakarta, informal settlements are hemmed in by new high-rise developments next to the priciest shopping mall in Indonesia (figure 3.2). As rural communities with communal ownership rights to land in China have been engulfed by urbanization, they have developed into warrens of multistorey accommodation with narrow alleyways. They are a prime example of how the 'differential value attached to what is

Figure 3.2 Informal housing behind Thamrin Square shopping mall, central Jakarta
Source: author

"formal" and what is "informal" creates the patchwork of ... space that is in turn the frontier of ... accumulation' (Roy, 2009b: 826). They are eyed greedily by both the state and real-estate companies. The question arises: what should communal 'owners' receive in the way of compensation if these 'slums' are demolished to make way for formal housing and commercial redevelopment? Individual householders are certainly keenly aware of the higher value of their piece of urban village, as the phenomenon of 'nail houses' (individual houses standing surrounded by sites cleared for new development) in China demonstrates. Original villagers might receive two apartments in a new development constructed on the same site or elsewhere in the city. But should villagers receive this much? Much of the increase in land value has come from urbanization in general and urban planning interventions. Moreover, in these urban village 'clubs', the legacy lingers of a socialist system in which all are equal but some are 'more equal than others'. Long-term migrant workers do not benefit at all, while village leaders may benefit disproportionately.

Housing as externality: club solutions to state failings

Yet housing can be viewed as an externality in grassroots NIMBY or 'build absolutely nothing anywhere near anyone' (BANANA) attitudes found across the global north. Frequently and knowingly these views elide protection of the environment with protection of private amenity and property values. The problem is two-fold, in that urban planning systems come to reflect established societal distributions of power and fail to properly price private access to public amenities such as green belts.

The housing question that urban planning has to deal with in wealthy countries has evolved from one of mass needs for shelter to grand desires for homes. Estimates suggest that prices can be anywhere between two and eight times higher than they might be without planning restrictions in and around London and New York (OECD, 2017b: 45). Planning restrictions – as restrictions on the supply of developable land – certainly contribute to house-price increases in contexts of rising demand as a result of growth in household formation (from ageing, single-person households and immigration), though they are by no means the sole cause of these increases. Moreover, housing shortages and rising house prices coexist with significant levels of inoccupancy. By the second decade of the current millennium, there were an estimated 1 million empty houses in the UK (Empty Homes, 2013, cited in Madanipour, 2017: 55).

In the US, privately developed master-planned communities can be considered clubs in search of a population. Some such as Reston in Virginia have gained their population only very gradually. Smaller, partial versions of new towns appear in the form of gated residential communities that provide some of the amenities and services typically found in the city. These are communities in which the rules of participation are predetermined and to which residents sign up in the form of home-owners' associations. These developments mark out other housing and other people as an externality – a financial burden, a criminal cost or an environmental impact – from which to shrink. Gated residential communities are clubs that exist within the inclusive clubs that are our cities. As many as eight different types of gated community have been identified in Argentina (Roitman and Phelps, 2011), with ambiguous relationships to the cities that host them. Gated residential communities may appear similar to cohousing projects such as those produced under the *Baugruppen* concept (building groups in which people join together to be the developer of their own separately titled homes) centred on achieving a strong measure of ecological sustainability and sense of community. The cohousing movement has its roots in experiments in Scandinavia in the 1960s and provides an alternative within the capitalist economic system (Sargisson, 2018: 145). The main difference between them and gated residential communities is the ends to which these particular clubs have been turned and the extent to which citizens are enabled to have a global sense of place.

The sorts of impacts that club solutions to housing provision make upon urban planning practice are unevenly felt across the globe. Secession into smaller and more particular club realms is a significant issue in some country contexts – notably the US. Across Europe, the issue is one of balancing the revenue benefits to municipal government as a club itself and the social exclusion costs of clubs existing within the club of the city. Across Asia, Latin America and Africa, the issue is whether the absence of urban planning (or its liberalization to facilitate housing for the growing middle classes) will further fragment already perilously fragmented cities (Webster, 2002: 400).

House as home: the latent informality of household desires

Shelter is not just a basic need; it is a fundamental site of urban planning – notably by citizens. It is the nexus of our being in a world with others. As an object, a house is a material part of the urban environment but, as a carrier of identity, a home describes a relationship between people and their environment (Dovey, 1985: 34). So it is that shelter provides few stable universal bases for urban planning by clubs or states.

The self-provisioning of shelter reveals that 'the performance of housing, i.e. what it *does* for people, is not described by housing standards, i.e. what it *is*, materially speaking' (Turner, 1976: 61).[1] The link between the functionality and affective properties of informal housing reveals how the desires that drive demands for urban modernity across the global south must not be ignored in urban planning thought and practice (Ferguson, 2006). The informal housing solutions found in cities of the global south reflect the desire to improve living standards through access to jobs, services and opportunities in the city. Here, informally developed suburbs are invested with a sense of an *approach to the city* rather than the escape from it assumed in Anglo-American reference points (Fishman, 1987). Finally, before we assume too much of the inherent advantages of the 'supportive shack' compared to the 'oppressive house' (Turner, 1976), the choice of informality is never one that is unconstrained (Ward, 2012).

Informality has been both explicitly and implicitly sanctioned in Canada and the US. Canada operated a 'build your own home' programme from 1942 to 1975 (Harris, 2004). In the US, over 500,000 people are now estimated to live in global south conditions close to the US–Mexico border, where lessons might be drawn regarding the balance of planning regulation and the feasibility and affordability of home ownership (Ward, 2012: 305). Otherwise, the desires bound up with the house as home remain a latent feature of urbanization in the global north. Here, there is something of a paradox regarding the travels of urban planning ideas, since, 'while ideas and policies espousing self-help originally evolved in developed societies and were exported with alacrity to developing countries, the relevance and applicability of such ideas for developed societies today are usually ignored' (Ward, 2012: 284). The informality of the UK's self-built 'arcadia for all' housing (Hardy and Ward, 1984) – scattered along roads and hillsides – was a powerful expression of house as home, yet prompted the modern planning system that banished it. The irrepressible desire to improvise is evident today in some of the most formalized urban planning systems in the world. Increased crowding into London has proliferated 'granny annexes' and 'beds in sheds' (Curran, 2012) and even the subdivision of individual rooms common in China. New 'McMansion' desires pop up with suburban 'teardowns' (Knox, 2008) and 'studentification' of neighbourhoods adjacent to universities (Smith and Holt, 2007). In Corviale, a 1960s state-built modernist social housing complex on the edge of Rome, self-built apartments colonize an entire floor intended as a shopping street. In Madrid, 'autoconstructed' housing such as that at UVA Hortaleza has been formalized by state-funded improvement and upgrading schemes (Tarazona Vento, 2018).

Health

Health and sanitation have provided the impetus to collective urban planning efforts dating back to ancient times (Ben-Joseph, 2012), but they came to the fore in early modern urban planning efforts directed at reforming the industrial city with its heavily polluted air and water sources and courses (Corburn, 2012). Critics of capitalism and social reformers documented the squalid and inequitable conditions in industrializing cities, helping to ensure the major water and sewer engineering projects and housing and urban planning reforms that proved so vital to the continued productivity of the same urban societies. The city is the best expression of there being shared positive 'externalities' – benefits to coexisting in close proximity. However, the intensive settlement of land produces accompanying negative externalities, where one individual's actions or one particular use of urban land has negative effects on neighbours. These negative externalities might be thought of as the pollution that accompanies city living. The management of these externalities has been at the heart of urban planning. In the modern era from the 1800s and at the city-wide scale, much urban planning knowledge and practice sought to regulate or zone land use and separate incompatible uses of land. The value of urban planning to the health of populations was measured in the productivity of urban industrial workforces. These basic concerns remain important across the global south in the enormous need for solutions to many of the same problems.

Speaking to the global north, Corburn (2004: 541) has suggested that 'urban planning practice shows few signs of returning to one of its original missions of addressing the health of the least well-off'. Modern health challenges have typically evolved to pertain to a much wider range of issues, including the susceptibility of burgeoning urban populations to pandemics, rising rates of obesity and mental illness, and the desirability of paying greater attention to the youngest and oldest in urban societies (Tewdwr-Jones, 2017).

Water, food and air

'Water has been an organizational and technological telos for the rationalization of urban space' (Gandy, 2014: 2), and this can be seen in the engineering projects of, for example, the Incan irrigation canals in Latin America and the aqueducts and sewers built for the ancient city of Rome. It is incredible to think that similar feats of engineering were not contemplated again until fully 2,000 years later, when Joseph Bazalgette designed the sewers of London. 'Though we can find metabolic continuities with the past, in terms of basic human needs the scale and complexity of contemporary urbanization involve a different order of dynamics and interdependencies' (Gandy, 2014: 10).

Indeed, water and the associated possibilities of energy generation provide some of the most massive examples of building projects. As often as not, they have been triumphs of engineering ingenuity over urban planning wisdom which have seen cities such as Los Angeles exceed the limits of their natural hinterlands. Issues of water management extend beyond cities into the water catchments in which they are placed and provide some of the best evidence of the enduring need to 'design with nature' (McHarg, 1969). Some of these catchments, such as the Murray–Darling Basin in Australia, are vast, cross state jurisdictions and have implications for agriculture, wildlife and city populations. The management plans required can be extensive and fraught with conflicts stemming from the different user communities. The construction of reservoirs almost always requires involuntary resettlement. Major concerns were voiced regarding the long-term impacts on livelihoods with the construction of the massive Three Gorges project to dam the Yangtze river in China (Wilmsen et al., 2011). Later investigations suggest that the close linking of the dam's construction to longer-term local development goals has begun to result in improvements in livelihoods (Wilmsen, 2016).

Clean water is vital to life but is scarce in many cities. Indeed, trends in national distributions of population reveal that we are moving in the wrong direction in this respect. The issue of water came to a head in Cape Town's 'day zero' in 2018, when drought conditions saw authorities announce that the city would run out of water in under three months. The crisis prompted changes in regulation, technologies and consumption by residents and businesses.[2] Water sources can become contaminated by wastes of all sorts in global south cities. While access to a basic need like clean water is often most apparent in informal settlements, stark differences – including the criminalization of self-help activities – exist across official and non-recognized slums in countries such as India (Subbaraman et al., 2012).

'Among the basic essentials for life – air, water, shelter and food – planners have traditionally addressed them all with the conspicuous exception of food' (Morgan, 2009: 341). And yet, of course, the food system has effects on other systems – public health, social justice, energy, water, land, transport and even economic development. The production of food is considered an essentially rural activity in planning systems that are typically based on sharp divides between town and country. The increasing use of land associated with rising urban populations is often treated as a local urban planning issue, but it presents a dilemma, since greater use of land may be essential to fulfil local basic needs in the short run but contribute to global ecological impacts in the long run. Part of the problem is that cities have become progressively divorced from their immediate environs and their original bases in nature, as registered in the massive infrastructures and logistical systems that deliver provisions to our doorsteps (Wackernagel et al., 2019). While urban planning systems seek to make the city more sustainable through the likes of green building codes and recycling schemes, 'business-as-usual' developments have seen the ecological footprints of most cities expand enormously in hinterlands that can be international in scope.

Agriculture never disappeared from cities and will need to experience a revival if extensive patterns of urbanization are to be sustainable in terms of food supply. The potential of urban porosity to supply some portion of the food requirements of city populations is signalled in the fine-grained mixing of buildings and garden allotments found across Tokyo. Otherwise, the emergence of urban farms and community gardening experiments seem likely to replace a proportion of the food bowls of cities being lost to the fresh allocation of land for housing at the urban edge. The Australian city of Melbourne's prime agricultural land is where much of the city's expected doubling of population from 4 million to 8 million will take place over the next decades. If the footprint of the city continues to extend as it has to date, the capacity of its food bowl to meet the city's needs could drop to 18 per cent by 2050 (Carey et al., 2016).

Images of air pollution in China's cities are familiar to us. Although one report found that twenty-four of the fifty most polluted cities worldwide in 2018 were to be found in China, it was India that had the seven most polluted cities in the world and seven more that featured in the top fifty.[3] The problem is not confined to rapidly urbanizing or global south countries, as, despite the implementation of congestion charging in London, it is estimated that nearly 10,000 people die prematurely as a result of air pollution in that city.[4]

The infirmity of the city?

In an age of greater than ever personal and virtual mobility, communicable diseases, sloth and old age have become urban health time bombs. New health concerns related to sedentary lifestyles have emerged and call forth the need for new planning imaginations to address issues of rising rates of obesity (Barton, 2009; Lake and Townshend, 2006). Some of the connections between urban planning and health are outlined by Giles-Corti et al. (2016), though contemporary compact-city ideals (of high-density, mixed-use developments in a walkable or public-transit-rich setting) greatly simplify the complexity of the connections involved. Moreover, the social determinants of health ensure that it is questions of justice and equity rather than physical urban form that have a strong bearing on the health of urban populations (Marmot et al., 2008). The World Health Organization and UN-Habitat Report (2010) *Hidden Cities* describes how health inequalities are 'systemic, socially produced (and therefore modifiable), and unfair'. They manifest in the form of food deserts where entire parts of cities have little access to fresh food.

In some global north societies, attention has turned to thinking about how the house can cater for needs across the family life course, including ageing populations. A scheme in the Admiralty

district of Singapore deliberately seeks to co-locate elderly populations with families with young children. In Newcastle-upon-Tyne
in the UK, the Future Homes Alliance scheme aims to provide
sixty-three homes suited to all ages and walks of life; it comprises
representatives of architects, house builders, energy companies,
social housing providers, ageing-interest voluntary groups, universities and digital companies working collaboratively to design and
build on a city-centre site. Such innovation has tested the limits of the
statutory system and its planning imagination, necessitating distinct
governance arrangements.

As I write in August 2020, the streets of Melbourne and most of the
world's great cities lie eerily quiet. The highly contagious COVID-19
virus reminds us all that what is most special about our cities – their
sociability – can also be their weakness. Living in close proximity in
cities makes the world population particularly susceptible to communicable diseases. These susceptibilities of cities have grown in scale
and complexity in a world more integrated in terms of total volumes
of people moving across borders. Infection can spread rapidly before
sequences of contacts, let alone origins, can be detected – as the
earlier SARS (severe acute respiratory syndrome) outbreak in 2003
had already indicated (Ali and Keil, 2006).

Mobility

If cities are evidence of the collective benefits of coming together
in geographical space, the irony is that cities are also notorious for
congestion. Urban planning interventions are not determined by
technological advances, but urban planning thought and practice
have more often than not reacted to the evolving technologies of
mobility. These technologies have *systemic* effects on the contexts in
which national and subnational planning systems and cultures are
nested. Nevertheless, different usage of modes of travel in notionally
similar cities suggests windows of opportunity are open for the urban
planning imagination to work upon (Stone, 2014). Urban planning
with and for automobility remains, but the emergence of new modes
of travel (such as electric scooters and guided electric vehicles) and
the resurgence of planning for pedestrians and cyclists mean that
negotiating our sense of being in the world with others will be
played out with respect to mobility choices on the city's streets and
pavements. Each technology of physical mobility has shaped not only

the geographical pattern of accessibility in and beyond cities but also, therefore, development potentials and land prices. As a result, though urban planning has often struggled to react to and form the complex ramifications of different technologies of mobility, this is also often the substantive concern over which there has been the greatest integration of practice (ESPON, 2018: 33).

While much of the discussion relating to climate change and sustainable development centres on cities, other, ostensibly inter-city, markets for mobility – long-distance, freight and air travel – are increasing and there is less scope to form them (Banister and Anable, 2009). Large changes in the pricing involved would be needed to shape these demands, for freight and air travel have become woven into global logistics corridors subject to some of the most exacting international urban planning approaches (Cowen, 2014).

Planning for mobility over the past century or so has had major consequences for the environment, for geographical patterns of socio-economic inequality and for the emergence of a 'non-place urban realm' (Webber, 1963). The urban planning imagination will have to come to terms with these. Will urban and transport planning do little more than cater to elite 'globals' with extremely high levels of personal physical and virtual mobility and a corresponding lack of attachment to place? What more can and should be done for those who have fewer mobility options and are highly reliant on local amenity and services?

Automobility and its systemic urban planning legacies

John Urry (2008: 344) describes how 'social life came to be locked into the mode of automobility'. Worse, urban and transport planning orthodoxy – 'predict and provide' – emerged to reinforce these configurations and has only recently begun to be challenged. The 'system' of 'automobility' can be considered an important ingre-dient in processes and patterns of globalization.[5] The numbers are staggering: 1 billion cars were made in the twentieth century, with 600 million of them remaining by the early 2000s. Car-based travel was expected to triple between 1990 and 2050. Vast spaces – roads and parking structures – are turned over to the car: about one quarter of the land in London and nearly one half of that in Los Angeles. For some, these are non-spaces (Augé, 1995), and they constitute a major challenge to urban planners because the logic of automobility appears to be the opposite of what we have to date considered to be urbanity. Where cities are intense concentrations of people in space, automobility promotes extension of human habitats (Sheller and Urry, 2000: 742). In particular, 'the urban environment, built during the latter half of the 20th century for the convenience of the car, has "unbundled" territorialities of home, work, business, and leisure' (Urry, 2008: 344), reinforcing the urban planning orthodoxy of separating 'incompatible' land uses. Little wonder that Rayner Banham (1971: 195) was able to comment how, by the 1960s in Los Angeles, 'the freeway system in its totality is now a single comprehensible place, a coherent state of mind, a complete way of life'.

Car makers and a highly active road-building lobby were able deliberately to undermine the viability of local public transit systems and continue to influence federal, state and local urban and transport policy in the US. In a pattern familiar in major infrastructure investments historically, the impact was felt predominantly by poor or marginalized populations and could have strong racial overtones, as in the complete destruction of Miami's black Overtown district (Rose and Mohl, 2012). In contrast, European cities retain tram systems or have restricted car movements in historic cores. Road building by US federal and state governments from the 1950s in the form of interstate highways, orbital beltways and parkways created major points of increased accessibility at the urban edges, turning agricultural land into land prized for the development of residences, business parks and regional shopping centres. Even the 'edge cities' (Garreau, 1991) that resulted now appear to be a thing of the past, as businesses and settlements become strung out along freeways (Angel and Blei, 2016). The distortions of automobility continue to unwind and they are obscured by tax and other incentives. The suburban residential dream continues to be peddled in the US and Australia, but the real costs of suburban living for poorer populations are revealed once the cost of ownership and use of cars to access employment is added to the lower prices of suburban housing. Awareness of 'transit deserts' has risen in cities such as Chicago, where NGOs have created tools to help calculate the real affordability of residential choices.[6]

High-speed rail and aeromobility: the interplaces yet to come?

Major increases in accessibility or time-space compression have been wrought by high-speed rail in the EU (Vickerman et al., 1999) and China (Chen, 2012). In China, high-speed rail may usher in major shifts in economic activity from historic to new peripheral centres. Similarly, the airport city or 'aerotropolis' planning model (Kasarda and Lindsay, 2011) presents a picture of the peripheral-ization of central business districts. Along with the unfolding effects of automobility, both these developments present the prospect of new 'interplaces' (Phelps, 2017) that the urban planning imagination will need to attend to.

The systemic properties of air travel are highlighted in the term 'aeromobility' (Cwerner, 2009). Here, then, 'in the global "space of flows", airports are critical nodes and have latterly assumed major economic significance extending beyond core aviation functions' (Freestone et al., 2006: 491). An emerging issue, as with many ports – sea, rail and air – is that these critical pieces of infrastructure are often removed from wider efforts at planning transport systems. The plans prepared by many airport authorities often sit uncomfortably with broader city- or metropolitan-wide land-use and spatial plans. Freestone, Williams and Borden (2006: 492) note the 'implications for spatial structure, transportation, commercial property markets, the environment and the efficacy of planning systems' of airport master plans. In Manchester, in the UK, the city's airport has been designated an enterprise zone, devoid of the usual urban planning controls, to encourage further airport-ancillary development.

Geopolitics intrude where there are concerns over foreign influence within logistics corridors – as with China's BRI. Otherwise politics has long surrounded the construction of airports. The machinations over providing for aeromobility in the UK are legend. Public inquiries on the building of London airport capacity date to the 1960s and have been considered an example of a 'great planning disaster' (Hall, 1980). Debate over the siting of an entirely new airport in the Thames Estuary continued up to the eventual decision, made in 2017, to expand Heathrow with an additional runway, which, at the time of writing, had been declared unlawful. In the process, individual planning methods (such as cost–benefit analysis) and planning itself have been the subject of ridicule.

Virtual mobility and its effects

Auto-, aero- and virtual mobility have facilitated lives that are now divided into fragments of time which are juggled by citizens. The upshot is that virtually all of us multitask, according to one survey, and this adds nearly 50 per cent more time in a typical working day (Kenyon and Lyons, 2007: 168). Our extended working day is now distributed across a variety of 'third space' venues (cafés, motorway service stations, co-working spaces, modes of transport) in different locations across the city rather than in the single office of an employer or client.

The urban plan for any given city may become a 'digital twin' of it, where the collection of real-time data via embedded sensors permits the intense monitoring of buildings, infrastructure and neighbourhoods. Originally pioneered to improve the performance of manufacturing systems, the concept has been deployed in the planning of Amaravati, the new capital city of the Indian state of Andhra Pradesh, which may be the first city born with a three-dimensional digital twin. The digital twin idea is closely associated with some of the methods I discuss in chapter 5 and raises issues of transparency, security and trust that have surrounded the making of plans historically. While they may enhance user satisfaction and the efficiency and profitability of urban assets, it is less clear how they could respond to issues of social justice.

Where the likes of auto- and aeromobility seemingly present the potential for the dissolution of the city, it is paradoxical that virtual mobility's effects may selectively help preserve the appeal of the oldest and more urbane of cities, since virtual mobility is most demanded in and can be overlain on some of the buzz of face-to-face contacts these cities remain famous for. However, it is here, in the likes of New York City, that the impossibility of top-down statutory urban planning to retrofit buildings and combine the efficiencies of ICTs with human creativity is found.

Sustainability

'The planning idea of analysing land-use change and considering alternative futures has always been conceptualised around the idea that human use and the management of land interact with natural processes and environment' (OECD, 2017b: 43). If urban planning imagination has been lacking with respect to the environment, it is worth emphasizing the constraints that urban planning faces as an activity nested within broader local, national and international systems characterized by significant professionalization and institutionalization. Urban 'planners have not been initiators ... Engineers made it their goal to counter the forces of nature while planners and architects provided the designs and rationales that sustained the transformation' (Hein, 2018: 8). This makes it all the more surprising that today it is 'land-use planning that has acquired the political burden of reconciling development pressures with environmental concerns' (Owens and Cowell, 2011: 4) – though this might be taken as a reverse compliment to urban planning.

The concept of sustainability has become accepted as 'a sort of meta-planning agenda' (OECD, 2017b: 48). However, it remains less clear how to achieve sustainability in urban planning practice. Historically, the environmental basis of urban planning has figured in a suite of designations to protect the countryside. Today, these designations have proved an imperfect means of controlling unwanted development and have stifled the planning imagination. Contradictions between the emphasis on environmental limits associated with approaches to sustainable development on the one hand, and urban planning orthodoxies on the other, have come to the fore. Part of urban planning's appeal cannot be separated from the elevation of specific urban forms – the grid, the garden or compact city – to iconic status. Yet 'sustainable urban development runs counter to the principles of the compact city in one fundamental respect: the primacy of process over form' (Neuman, 2005: 11). Moreover, the sustainability agenda is nevertheless one in which there is a strong *outcome* ethic, as opposed to one focused on the procedural integrity designed to ensure communication or collaboration in an open-ended *process* (Owens and Cowell, 2011: 8). It is these contradictions that haunt urban planning's record in dealing with substantive concerns relating to the environment.

Environmental designations: the paradoxes of planning for the city and its other

There are paradoxes in the manner in which environments – built and natural – have been conserved, and these often stem from the influence wielded by the most powerful in societies. As recently as the 1960s there were proposals to demolish large parts of Georgian- and Victorian-era London, but today these and older built environments in London and many other cities worldwide are well protected by conservation areas, zoning overlays and other regulations. A remarkable revaluation of buildings of historic character has been made by middle-class gentrifiers and developers aware of the new economic potentials of residential and industrial heritage (Nasser, 2003), and by politicians and economic development planners attuned to tourism possibilities. Planning as urban conservation has preserved much of what is visually most appealing in many cityscapes, though it entails significant trade-offs with the afford-ability of housing and rigidities in commercial real-estate markets where land uses may remain well below best value – as captured in the juxtaposition of Melbourne's Victoria Market with the remainder of its downtown (figure 3.3).

National parks, designated in countries like the US and the UK to protect relative wildernesses, valued landscapes or fragile ecosystems, now suffer from an intense 'tourist gaze' (Urry, 1990) that threatens their very nature. Already enclosed in an attempt to protect against a tragedy of the commons that I discuss in the next chapter, their scarcity value generates a demand and a consequent need to restrict use.

Figure 3.3 Melbourne's Victoria Market and downtown
Source: author

Green-belt policy, pioneered in the UK and widely exported and adapted as urban growth boundaries and green wedges or fingers, provides insights into the city's relationship to its other: nature or the rural. Green belts can be vital in preserving valued natural landscapes, food production and the functioning of ecosystems essential to protecting and nurturing citizens. In the UK, the planning imagination regarding the green infrastructure or ecosystem services provided by green belts on behalf of urban populations became frozen at the point of designation – although practices in other European nations have a better record (Keil and Macdonald, 2016). For those who own homes within green belts, the policy is one that preserves private property values and private – not public – amenity. For local political representatives, green belts and other 'green-gap' planning policies are those on which they get re-elected. Urban planners working in the state sector often complete the circle of self-delusion when offering nothing more intellectually than a need to prevent the coalescence of settlements. Apart from the questionable natural qualities of some land included within green belts, the designation of green belts and growth boundaries can have multiple unintended consequences. These include the 'blight' that can exist inside green belts as the 'hope values' attached to agricultural land see it fall into disuse, and the production of non-contiguous, scattered, urban sprawl.

In the seemingly highly planned developmental state economies of East Asia, green-belt policies have never held strong. Tokyo's green belt, designated in 1939, disappeared under the inexorable westward spread of urban development in a booming national economy. In Beijing, urban planners hedged their bets by defining an inner and an outer green belt: the latter was needed because 80 per cent of the first green belt was cancelled due to having become urban, and yet over 40 per cent of the second green belt is now urbanized (Liu, 2008: 323). In the latter case, the value of green belts may only be recognized when it is too late. In the former case, the mixing of urban and rural land uses at a fine grain may offer an alternative to green belts (Yokohari et al., 2008). In those countries where environmental designations command popular, political and professional planning support, they are enforced, but their existence is intimately connected with issues of housing supply and affordability. Here, fears that large parts of countries have already been or will be built on are unfounded. Only 8 per cent of the land area of England is of 'developed use', with 40 per cent protected by green-belt, National Park, Area of Outstanding Natural Beauty or Site of Special Scientific Interest designations (Ministry of Housing, Communities and Local Government, 2019), the latter two designations having grown significantly during the 1980s and 1990s.

Eco- and other city forms

'There is ... widespread recognition that the spatial configuration of cities and the ways in which land is used and developed have significant implications for both adaptation to the adverse impacts of climate change and reduction of the emissions that are causing the change' (Davoudi et al., 2009: 13). Beyond this, however, agreement ceases regarding the contributions of competing urban forms to sustainable development. Quite apart from the compatibility of growth and sustainable development agendas (Alexander and Gleeson, 2019), there are a number of different ways in which population and employment growth can be accommodated, such as infill, urban extensions and entirely new settlements, each with its own economic, social and environmental costs and benefits (Green and Handley, 2009). It is worth reflecting on just how quickly the orthodoxy of urban decongestion associated with the garden-city idea was replaced in the new urban planning orthodoxy of compact and eco-cities. Yet this orthodoxy conceals as much as it reveals.

In some versions of the sustainability agenda, the environmental dysfunctionalities of extant cities are to be overcome with the building of entirely new eco-cities away from the old. In China, new eco-city building has been as likely to fail altogether (as in the case of Dongtan near Shanghai) as to materialize (as in the case of the Sino-Singapore Tianjin Eco-City (SSTEC) in Tianjin), regardless of the actual ecological credentials of the developments concerned. The SSTEC has innovative experiments including recycling schemes that generate local currency for residents (though not for migrant construction workers) at local shops, and residential properties in the SSTEC already command a premium, particularly luxury developments adjacent to golf courses – whose environmental consequences surely need no elaboration. Away from the modest numbers of these attempts to make ecological utopias, 'the greater planning challenge will be retrofitting existing cities and urban areas' (Beatley, 2012: 120).

Urban metabolism in an urban age of climate change and hazards

Our Common Future defined sustainability as that which 'meets the needs of the present without compromising the ability of future generations to meet their own needs' (World Commission on Environment and Development, 1987: 8). The notion of sustainability foregrounds the near-inevitability of a 'tragedy of the commons', discussed in the following chapter, and reinvigorates ecological resonances of urban planning, but now in terms of limits and thresholds.

Beyond limits, some effects can be irreversible. Yet the notion of limits is at odds with the wickedness of the sustainability problem, which has seen it defined in an inclusive way to cover environmental, social and economic sustainability without a clear hierarchy among these three. Political and policy trade-offs are inevitable and strong local popular and political coalitions would be needed to ensure that either social or environmental concerns offset the economic tendency of 'business as usual' (Rydin, 2009). Thus, 'planners are being asked to reconcile, trade, and indeed, overturn short-term and long-term expectations for development' (Davoudi et al., 2009: 7), though for some 'the emphasis on the integrative and intra-generational aspects of sustainable development has obscured that of inter-generational justice and long-term horizons' (Wilson, 2009: 224).

'The focus on ... living and designing within limits ... along with the urgency of change remains an important mental framework for planners' (Beatley, 2012: 93). Indeed, the focus on limits in a context of the pervasive environmental effects of climate change renders some of urban planning's traditional concerns quaint: 'in the past, large scale pollution from point sources such as factories had a significant environmental impact. Today, steady, incremental, and accumulated pollution and resource depletion from nonpoint sources that emanate from the daily and routine actions of all have significant repercussions' (Neuman, 2005: 16). With the separation of incompatible land uses now well established and enforced as an urban planning principle, a focus on limits has come to the fore in carbon-neutrality and/or zero-carbon demonstration projects, and in building standards found in the global north. In the global south, the coming together of both these types of externalities has simply overwhelmed planners in a context where plans have little effect on actual development patterns and where they cannot be enforced.

Only two global north countries appear in the list of the top twenty most vulnerable to climate change: Australia and the US (Halsnæs and Laursen, 2009). Although the numbers of casualties from climate-related events in the global north can be significant – 20,000 died in European heatwaves of 2003, and 1,101 as a result of Hurricane Katrina in the US in 2005 (Davoudi et al., 2009) – these figures are dwarfed by those in the global south. Each year, 200 million people are affected by climate-change-related events (Halsnæs and Laursen, 2009), the majority of whom live in the global south. Within the global south, the problems of climate-change vulnerability, mitigation and adaptation often come together in hot and humid climates, which drive energy demand in the form of air conditioning where wealth allows, but where flooding and storm damage affect many poorer populations (Pizarro, 2009). It is estimated that in more than one half of the world's coastal countries, at least 80 per cent of populations live within 100 km of the coastline (Blackburn and Marques, 2013). Coastal urbanization has increased in global south countries, where 'the unique juxtaposition between large scale settlement and environmental processes that exists in coastal megacities ... culminates in heightened risk' (Blackburn and Marques, 2013: 13).

Cities across the vast archipelago nation of Indonesia exemplify the risks to coastal settlement, whether from 'infrequent' catastrophic events, such as the 2007 Indian Ocean tsunami that devasted the province of Aceh on Sumatra, or the regular problems of tidal flooding that affect all of the cities on the northern coast of Java. Jakarta is a prime case of the latter risk: 40 per cent of the entire Jakarta metro area is below sea level, with the most severe recent tidal flood in 2007 displacing 422,000 people (Pelling and Blackburn, 2013: 202). Indeed, urbanization has compounded the problems, since development of the coastline in and around Jakarta has destroyed 80 per cent of the residual mangrove cover that existed fifty years ago – cover vital as a natural first line of defence against flooding events. In cities along the northern coast of Java, tidal floods combine with disposal of waste to create a toxic problem. In Jakarta, approximately 20 per cent of garbage generated each day is disposed of into rivers and drains or burnt by households (Pelling and Blackburn, 2013: 201). In Pekalongan, the toxic dyes used to colour batik cloth now colour the water and jeopardize the livelihoods of fishing villagers.

The state of California is home to extensive patterns of urbanization, and yet it has also been in the vanguard within the US in enacting environmental legislation. State-wide targets for reductions of greenhouse gas emissions have been more ambitious than those set in the Kyoto Protocols. The state has pioneered 'backcasting' techniques and generated a strong measure of inter-agency cooperation and stakeholder involvement. Indeed, 'never before has this style of planning been attempted on a scale that will require change in virtually every aspect of economy and society. California's climate change planning may thus be seen as a significant step towards a new style of planning appropriate to sustainable development generally' (Wheeler, 2009: 131). And yet 'Californians do not seem ready to question ... basic elements of their lifestyles' (Wheeler, 2009: 134). California and other examples speak to the idea that local discretion or experimentation is key to generating a variety of place-sensitive responses to more than local issues, notably where there are policy vacuums at the national level.

'Concentration of population in cityscapes dominated by technology and built infrastructure has fostered the conception of an urban society that is increasingly decoupled and independent from ecosystems' (Gomez-Baggethun and Barton, 2013: 235). The decoupling here is inherent in some urban planning systems and the planning imagination's inability to come to terms with in-between landscapes. 'Our move toward sustainable cities will require an important shift in thinking about cities not as linear resource-extracting machines but as complex metabolic systems with flows and cycles' (Beatley, 2012: 110). There remain problems in valuing urban ecosystems characterized by the complexity, heterogeneity and fragmentation of cemeteries, yards, allotments, urban forests, wetlands, rivers, lakes and ponds, with functions that encompass food supply, water flow and runoff regulation, temperature regulation, noise reduction, air purification, pollination, seed dispersal and animal habitats (Gomez-Baggethun and Barton, 2013: 236). Urban planning can seek to take account of ecosystems by raising awareness, setting priorities and providing incentives (Gomez-Baggethun and Barton, 2013: 241). Something as simple as raising awareness changes land and property valuations. However, more often, the valuing of ecosystem services sits uncomfortably with politically least-sensitive 'escape routes' for development and monetary-based valuations, as in the UK (Harris and Tewdwr-Jones, 2010).

Singapore provides a case of planning for the biophilic city: 'Although Singapore's verdance is partly a function of a tropical environment in which everything seems to grow well, there is also much conscious intervention here' (Beatley, 2016: 52). It is now home to several 'vertical gardens' – for example, the thirty-six-storey residential tower of Newton Suites Solaris, and the Park Royal Pickering Hotel. These are green innovations which, incidentally, generate business tourism. Away from the tourist gaze, renaturing of concrete storm-water channels has taken place in Bishan (figure 3.4). Rejuvenation of river courses has been one of the most remarkable recent reverses in city building, given the enormous energy and financial resources put into channelling, burying and diverting rivers during the twentieth century; though in some instances – such as the recovery of the Cheonge Cheon river from under a highway in central Seoul – the practice may be little more than a gesture motivated by a desire to mobilize the rediscovered connection of urban amenity and economic growth.

Localized experiments in sustainability are a means through which policies diffuse and of knowing and managing cities (Bulkely and Castan Broto, 2013: 367). Experiments may contribute to the

Figure 3.4 Renaturing of a storm channel in Bishan, Singapore
Source: author

conflict and debate necessary for robust planning processes of the sort I discuss in chapter 5. They may also be an important way in which new mixes of the strengths of different urban planning actors are brought into being.

Economy

Urban planning takes place increasingly in some variety of a capitalist economy. Yet it is a capitalist economy that has its alternatives, which will be an ever-more important reference point for the planning imagination.

Economic growth: a state obsession

Official measures of economic performance, of which gross national product and gross domestic product are the prime examples, reflect a series of national obsessions with growth which find expression at the urban scale. Gross value added and other measures which provide indicators of economic development (improvements in productivity and different ways of producing) have a part to play in reframing urban planning.

The separation of land uses at the heart of modern urban planning systems has had unwanted effects on the economic sustainability of cities, not least in terms of entrepreneurship but also in the variety of encounters that are said to be the signature of inclusive cities (Fincher and Iveson, 2008). The home has re-emerged as a location of enterprise, but urban planning has often failed to adapt to these realities. Folmer and Risselada (2013) identify the importance of residential suburbs to the formal economies of Dutch metropolitan areas. Different planning approaches can produce different outcomes in terms of rates of entrepreneurship, leading these authors to suggest that excessive regulation of neighbourhoods can produce poor outcomes in terms of entrepreneurship. Some of the most pernicious of these regulatory effects of urban planning affect those locales and populations where the possibilities presented by informal enterprise are most needed. Where modern urban planning thought and practice have sought more systematically to embrace these developments, it has been in the form of temporary exceptions to predominant land uses in the form of pop-up events, markets and urban farms. In Johannesburg, the peripheral township economy has been recognized by the city's planners, and steps are being made to provide infrastructure in so-called 'Corridors of Freedom' (Harrison et al., 2019) that link these areas and sectors to advanced services downtown via rapid transit systems. To ignore these employment spaces is to ignore a major part of the employment prospects across the global south.

If cities are creative triumphs (Glaeser, 2011), then it is also true that the decline and disappearance of cities is a matter of historical record that gnaws at the heart of planning. It gnaws because so much of the economic activity of cities is unthinking 'business as usual' – and that business is the business of producing more by using more. Moreover, some of the more conspicuous instances of the creativity and inventiveness of the city are extremely fragile. The likes of creative quarters usually appear spontaneously as a product of the 'rent gaps' arising from rigidities in urban land and property markets. They can be further stimulated by planning policies that facilitate the warehouse conversions and 'loft living' apparent in many cities. The problem for creative and cultural industry enterprises and for urban planners is that 'cities are wasting what are currently one of the most dynamic industries as "starter fuel" for property development and residential expansion' (Pratt, 2009: 1043). Beijing's 798 art district has largely transformed from the production to the retail of art and cafés and restaurants (Currier, 2008). Seoul's Insadong area has become less distinctive and more corporatized (Douglass, 2016). The problems associated with a similar over-politicization and over-planning or corporatization of London's Tech City were crystallized with the arrival of Google in an urban economy noted for its start-ups (Nathan and Vandore, 2014).

The circular urban economy

'The capacity to make, use, and discard large quantities of items became the ... hallmark of city life from the very beginning', to such an extent that excessive consumption and waste are 'not a *modern* problem but an *urban condition*' (Smith, 2019: 108, 156, original emphasis). This urban condition ensures a circular economy potential of cities.

> In highly developed economies of the future, it is probable that cities will become huge, rich and diverse mines of raw materials ... The largest,

most prosperous cities will be the richest, the most easily worked, and the most inexhaustible mines. Cities that take the lead in reclaiming their own wastes will have high rates of related development work. (Jacobs, 1969: 110–11)

Local currencies have been developed globally to prevent 'leakages' (non-local expenditures) and generate economic circularity or local economic cohesion (Williams, 1996). Historically, such local currencies have also been turned to dark ends by mining companies that practised model town planning but wished to minimize labour turnover.

Broader notions of self-sufficiency have emerged in the 'transition towns' movement, which numbered 714 separate initiatives found in small towns across thirty-one countries (Taylor, 2012). The challenge, here, as Taylor (2012) notes, is how numerous local initiatives relate to the major city economies. Here the notion of the circular economy is one that anticipates a decoupling of economic growth from natural resource depletion and environmental degradation. It has been further codified in the RESOLVE framework of the Ellen MacArthur Foundation.[7] However, as applied to the economies of cities, RESOLVE has limitations which reveal the wickedness of urban planning problems. In particular, following Williams (2019), RESOLVE is based primarily on fostering circularity within a purely economic system, not an urban ecological system. RESOLVE focuses on production rather than consumption, despite the fact that many global north urban economies are by now oriented not to needs but to the wants of consumption, and production has shifted to the in-between spaces I discuss in the next chapter. The RESOLVE framework ignores land – at once the most important and yet lumpy and inertia-prone of resources in the city, and which urban planning seeks to operate on. Land is the source of extraordinary 'unearned' wealth, as we will see in the next chapter: wealth that continues to drive urban planning by individuals, clubs and even states in ways that have little regard for social justice or environmental sustainability. The RESOLVE framework ignores the inertia present in fixed infrastructures that are part of the city. These have been developed largely in parallel with one another, with limited interchangeability or adaptability, and consume vast amounts of resources in their maintenance and replacement. Finally, circular economy ideas that are codified in RESOLVE do not address the fundamentally multi-scalar nature of the urban planning challenges involved in a highly globalized urban economic system.

The alternative economies of clubs and citizens

The Greek root of 'economy', *oikonomia*, refers to the household, not the locality, the region, the nation or the global economy. It is in the home that tensions between needs and wants, and the sources of inspiration for how we balance them in urban planning designed to sustain the economy of cities, might be found. In the home, many different forms of work – paid and unpaid, capital-acquiring and resource-conserving – come into intimate relation, forcing us into important moral choices regarding the stuff we accumulate (Miller, 2010). Moreover, if extended-family living has been the norm for much of time and across much of the world (Gibson et al., 2018), then there are clear implications for the revision of urban planning systems geared towards the single-nuclear-family detached home of US suburbia and aspired to by affluent households, modernizing states across the global south and the clubs of gated communities everywhere.

The economy can be reframed around a host of alternative or non-capitalist economic relations that urban planning might seek to regulate less or incentivize (figure 3.5). The individuality associated with capitalist economic relations is only the tip of the iceberg and is more than matched by a moral, sharing or gift economy sometimes associated with clubs. Indeed, a review of the data on people's use of time for paid and unpaid work reveals the 'shallow penetration of capitalism' within the totality of economic exchanges across global north and south (White and Williams, 2018: 176). However, the issue as it pertains to urban planning imagination, policy and practice is how to recognize 'alternative' economic activities and – as with 'ecosystem services' – value them appropriately.

Across cities of the global south, statutory urban planning arrangements, such as separation of land uses, are insufficiently attuned to the fine-grained temporal and locational niches exploited in informal economic activities, which can constitute as much as 80 per cent of the business stock (Skinner and Watson, 2018). Model vendor relocation schemes such as in Solo, Indonesia (Phelps et al., 2014), risk substituting wrongly located and overdesigned formal markets for the 'natural' markets that have grown over time along streets. The right to work and the suitability of streets for that work are legally recognized in several global south countries, though tolerance of the likes of street vending has been hard fought in these same countries (Meneses-Reyes and Caballero-Juárez, 2014).

The informality that characterizes alternative economic forms in the global north is due in part to the inequalities and crises of the urban economic growth of capitalist economies. The possibilities for urban planning to recognize and support household-originated, grassroots or community economic initiatives as an alternative to the growth-maximizing tendencies of nation states rest on the fostering of a global sense of place. 'In a community economy we need to keep our eyes on the interdependencies between the different kinds of work we do' (Gibson-Graham et al., 2013: 38). Communities are increasingly woven into the global commons through alternative trade networks, such as those concerned with fair or ethical trade of commodities, which implicate consumers and producers in an ethic of care for distant others (Gibson-Graham et al., 2013: 103).

The building of community economies is not free from contradictions. Examples of sharing, gift-giving and moral economies are found not only in in alternative capitalist and non-capitalist but also in fully capitalist forms (Hall and Ince, 2018). An example of the last is the platform economy Airbnb, which mobilizes 'excess' residential capacity but which has exercised the minds of urban planners in many cities across the world, since the revenues it has generated have been capitalized into rising property prices. Sharing or gift economies may appear to imply sustained interpersonal interactions and the forging of the sorts of intersubjective understandings that communicative and collaborative planning processes seek to foster among urban planning actors (chapter 5), and yet they can, and often do, involve mediation via ICTs in a non-place urban realm (Hall and Ince, 2018). A geographical sensibility is vital to urban planning being able to recognize and liberate the potential of alternative and non-capitalist economic forms, since they are found across multiple and different spaces, ranging from the intimate surroundings of homes to the planned and conserved public spaces of cities, the vacant interstitial spaces of the city and the non-place urban realm of ICTs. Non-capitalist transactions are pervasive in capitalist economies of the global north. One reason for this is that 'things are of value when they meet our needs to collectively see through to the real and to increase variety and complexity', thereby

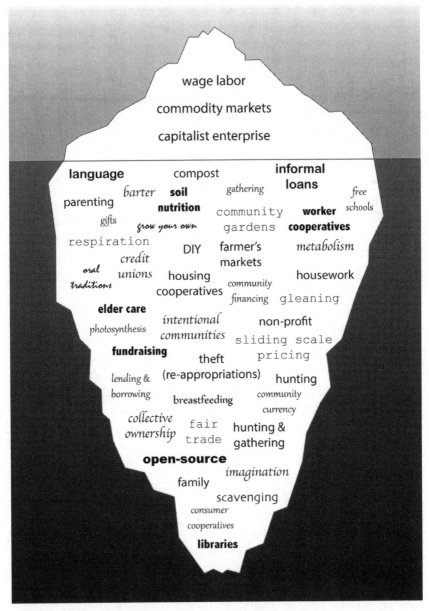

Figure 3.5　How to reframe economic development
Source: reproduced from Community Economies Collective, licensed under a Creative Commons Attribution-ShareAlike 4.0 International licence. https://www.communityeconomies.org/resources/diverse-economies-iceberg

'generating a form of wealth that can be called gift-value' (Sack, 2003: 256). The gift-giving economy is an open-ended spiral of greater value that is accessible to all (Sack, 2003: 263) – though it is apparent that gift-giving can become commodified (Miller, 2010).

Non-capitalist economies are thrown into sharp relief when considering the lives of indigenous peoples. The customary clan-based property relations of indigenous populations in Canada and Australia are different from those typically enshrined in the private property relations of modern urban planning (Blomley, 2014). Here we see how value traditionally inheres at least as much in the *relationships* associated with the circulation of things as in the things themselves, if not more so (Gombay, 2018). Restorative justice thus focuses on questions of 'the extent to which Indigenous peoples are able to shape economic practices ... whereby alternative models for conceiving of and enacting economies express Indigenous peoples' values' (Gombay, 2018: 161). In the case of the Plan Nunavik alternative to the Quebec government's Plan Nord, the emphasis is on the right to participate in new economic activities associated with large-scale resource extraction, with indigenous corporations having been established to safeguard the collective economic interests of beneficiaries to land claims established in a series of commissions organized by different tiers of government, while at the same time preserving traditional practices of subsistence (Gombay, 2018).

Conclusion

Different substantive concerns have come on and off the agenda in different parts of the world at different points in time. Some part of the shifting of priorities reflects not merely general societal development but the need to deal with those effects. Even seemingly separate substantive areas of concern for urban planning are closely interconnected and reveal the fundamental wickedness of the problems that urban planning is addressed to – a wisdom that I discuss in the next chapter.

The discussion of different substantive concerns hints at the different urban planning actors involved and the methods they deploy, but I discuss them in greater detail in chapter 5. In their variety, the methods themselves reveal the complex and open socio-environmental systems shaped by these innumerable acts of urban planning. From the desires that shape individual travel and home making, to the expert forecasting of aggregate patterns of travel, to the informed guesswork of scenario planning, they reveal the nature of urban planning as part art, part science. As we build our cities, the urban planning imagination will need to continue to search for productive mixes of the wisdom and methods distributed across citizens, clubs and states and associated with a variety of substantive concerns.

4 Wisdom: what does planning teach us?

Introduction

The urban planning imagination carries wisdom, and it would be well if as much of this wisdom as possible is remembered and renewed in the building of our cities. If the theory and practice of urban planning are a store of valuable abstract and practical learnings relating to the organization of human activity on earth, no single set of actors, place or community has a monopoly on that wisdom.

The history of urban planning provides a reflection on the society of the time and ourselves as planners – as citizens, as part of clubs and states (Gold, 1997). Urban planning *theory* has evolved over time, with the critique launched at physical planning orthodoxy held responsible, among other things, for the destruction of historic cores of cities and the perpetuation of inequalities. Urban planning *practice* continues to evolve into something more complex and self-consciously ethical. Within this apparently linear history of urban planning thought and practice, many of the achievements of planning present themselves in moments when an idea meets its time and place. We see some of this with the Radburn layout, discussed in chapter 2, which 'was the right development at the right time' (Birch, 1980: 426).

Yet the history of urban planning is not linear. 'There is no easy and reliable way by which in future we can avoid the errors we made in the past' (Hall, 1980: 267). Lessons from past urban planning thought and practice appear to have been forgotten and urban planning knowledge lost as part of 'institutional amnesia'. It may seem paradoxical, but loss of information will probably increase in an era of more information of all sorts (Pollitt, 2000). The challenges of discerning and retaining good information and of guarding against individual and collective amnesia in public policy formation have never been greater.

Justice and equity

'If there is one "ought" about the city, it is that it has to ensure that justice be done by its citizens ... not only done, it must be seen to be done' (Rykwert, 2000: 228). Recognition and rejection of injustice may be instinctual (Fainstein, 2010), but there is plenty of evidence to attest to the injustices of the city. The *World Inequality Report 2018* (Alvaredo et al., 2017) suggests that the share of US and Western European national incomes of those in the top 1 per cent of incomes rose inexorably and that of the bottom 50 per cent of incomes fell from 1980 to 2015. Inequalities have been perpetuated in comparatively modern beliefs which present it as natural or inevitable (Dorling, 2011). Across the global south, the quest by insurgent elites to accumulate capital in the form of land and real estate contributes simultaneously to the 'impoverishment of poverty' as a substantive concern (Bhan, 2019). It will be important for the urban planning imagination to continue to confront the pervasiveness of these modern beliefs, not least since they close off serious consideration of urban planning alternatives.

The urban planning associated in history with the production of downright evil places – such as the concentration camps of the Nazis – might be characterized as seeking excessive restriction, impermeability, isolation and autarky (Sack, 2003: 25). In everyday terms, poor places can be the product of the sort of unthinking 'business as usual' that characterizes the energies of some makers of the built environment – speculators, financiers, developers, constructors, commercial and household investors, property managers, architects and, of course, urban planners. Instead, then, because 'we continually change reality by realizing (some of) our imaginings, it is essential we think through whether these changes are good or bad' (Sack, 2003: 16). While the search for the good may be compelling, the vast majority, if not all, of the places we have fashioned as humans could be said to be morally mixed. From a normative perspective, good or better places are those that generate or provoke awareness on the part of their inhabitants and users and those that are varied and complex rather than simple and monotonous (Sack, 2003: 25). Awareness, variety and complexity in the production of good or better places evoke the relational or global sense of place that I introduced in chapter 2, since they refer to 'a world to see and points of view from which to see it' (Sack, 2003: 160).

Debate continues as to whether our global lot has improved, but a concern with social justice within urban planning thought and action is one that has continually to be worked at. Of course, a sense of (in)justice takes on different meanings depending on social, geographic and historical context (Harvey, 2000), and this sense can apparently go as quickly as it comes. In my lifetime, London has become glamorous in a way unrecognizable as recently as the 1980s of riots, physical decay and population decline. Yet the revitalization of London has occurred at the same time as differences in wealth and opportunity have increased, and at the same time as much of the sense of social justice inhering in the urban planning imagination has receded from view. One of the ingredients found in definitions of urban planning is the evaluation of alternatives. Alternatives to a particular proposal or a particular plan are one expression of the plurality of voices that must be heard in the shaping of our urban futures. Alternatives can provide a basic check on business-as-usual tendencies in the building of our cities, not least because physical environments describe a substantial part of the opportunities we draw on individually and collectively in the course of our lives. One person's conservation of the status quo is another person's opportunity lost (Rydin, 2011: 73). However, today it would seem that urban planning's instinct against injustice has met its moneyed nemesis in the form of TINA ('there is no alternative').

Urban planning educators may inadvertently have played a role in urban planning's retreat from questions of social justice, given the increasing emphasis on the processual integrity of urban planning or inputs to it, rather than its outcomes (Marcuse, 2012). The emphasis began to emerge in the 1970s in terms of operations research (Friend and Jessop, 1969) with its affinities to popular and political calls for planning processes to become more efficient. Concerns with the efficiency of planning – at the expense of social and spatial equity – have found strange bedfellows more recently with the attention paid to the proliferation of rankings and competitiveness indicators that now indirectly drive the politics informing urban planning agendas. However, as Fainstein (2010: 9) notes, 'the justice criterion does not necessarily negate efficiency and effectiveness as methods of choosing among alternatives, but rather requires the policy maker to ask, efficiency or effectiveness to what end?' Here, urban planning thought and practice can usefully turn attention to those 'strong' forms of inter-locality competition that emphasize the uniqueness of places rather than the 'weak' competition of imitation (Cox, 1995).

Emphasis shifted to the procedural justice of urban planning from the 1990s in theories of collaborative (Healey, 1997) or communicative planning (Forester, 1993; Innes and Booher, 2010), flowing from Jürgen Habermas' ideas of the potential of communicative action for the renewal of the public realm. If 'planning is about deciding whose voice should be heard in determining ... issues and, ultimately, whose voice should count' (Rydin, 2011: 10), then as socially concerned as some of its main promoters have been and as influential as this line of thought has been inside the academy, these theories de-emphasize the potential for distorted communication as part of the 'dark side' of existing power structures in society (Phelps and Tewdwr-Jones, 2000). They do so to the point where urban planning thought and actions may barely concern themselves with outcomes, let alone socially or spatially just outcomes (Fainstein, 2010: 3–4).

The age into which planning for the just city is pitched is one in which many of the unintended consequences of previous urban planning interventions have come to light, generating new knowledge, new political sensitivities and new planning sensibilities attuned to fine-grained societal differences and an individualization of politics (Beck, 2002). Where in the past the just city was signalled in social (re)distributive planning for relatively spatially equal access to universal services – schools, parks, hospitals, transit – consumed collectively by a 'universal citizen' (Fincher and Iveson, 2008), today these questions of redistribution cannot be separated from related questions of the recognition and representation of an expanding range of gender identities and sexual orientations and the ethnic superdiversity of cities. The challenges of these elements to urban planning thought and practice are revealed in the 'dynamic interplay of variables among an increased number of new, small and scattered, multiple-origin, transnationally connected, socio-economically differentiated and legally stratified immigrants' of superdiversity (Vertovec, 2007: 1024).

Much of the urban planning imagination might be thought to work on the assumption of at least the possibility of fair outcomes, but the lesson from polarized cities is the likelihood that ethnic groups will question the legitimacy of urban planning as systemically incapable of producing fair outcomes (Bollens, 2002), with implications for what collaborative, communicative or deliberative planning processes can achieve. The urban planning imagination's power to provide adequate theoretical frameworks, let alone practical guidance, for these senses of difference now exploding in societies has been overrun before it has been able to address more immediate questions such as what a non-sexist city would be like (Hayden, 1980), how to dismantle ingrained and systemic racism within government and other institutions shaping cities (Gillette, 2011; Thomas, 2000), or how to better plan for disabilities (Imrie, 1996).

If cities allow for 'aspirational consumption of all kinds and the exercise of multiple identities' (Smith, 2019: 260), then the uncertainty, risk and individualization of politics associated with superdiversity, entrenched polarization and expanding senses of difference found in cities push in the direction of constructing urban planning anew, with the purpose of leveraging the expertise and knowledge of a mix of actors – though the challenges remain formidable. The right to the city is the right to encounter (Fincher and Iveson, 2008). Urban planning can help affirm identities but that is not enough. Urban planning for a more just city needs to do more than cater for interaction or encounter per se; it almost certainly needs to combine recognition and redistribution to transform urban conditions. 'No encounter, planned or not, can be envisaged without some form of exclusion ... So one of the tools of planning for encounter is to make sure that forms of inclusion and exclusion that might arise ... are anticipated ... acknowledged and compensated for if they privilege certain groups and behaviours over others' (Fincher and Iveson, 2008: 170).

The disruptive potential of temporary urbanism was noted in chapter 2. One problem is that the reality of a lot of what is done in the name of diversity in cities by urban planning overlooks how the encounters promoted can be entirely oriented towards consumption. It is on these grounds that planning interventions in the form of events such as festivals or temporary uses of urban space have been criticized. Such criticisms of urban planning 'tactics' may be a little unfair, given what Fincher and Iveson (2008) describe as their paradoxical function: at once promoting interaction and cohesion and also acting as opportunities for subversion and protest. There continue to be vexed problems for urban planning to deal with here, as I hinted in the previous chapter regarding the dissolution of creative and cultural quarters in cities. While, 'in place of separation and confinement ... the inherent hybridity of both place and people ought to be acknowledged and embraced as a condition of urban life' (Fincher and Iveson, 2008: 149), in practice, for Stretton (1975), production requires the seclusion of the suburbs. Here, two just city concerns come face to face: the critique of much contemporary consumption, spectacle and mega-project-oriented development focused on central business districts, and the call for a greater emphasis on redistribution and production to focus on neighbour-hoods (Fainstein, 2010). The notion of community in a nurturing urban setting becomes rather one of a community of communities in which, to an extent, inequalities dissolve as individuals not only enjoy a direct identification with an individual place, but also move among and have variable associations with multiple other places (Fincher and Iveson, 2008: 87–8). The urban planning imagination will need to incorporate a sense of place in which consumption and production and rather different senses of community are brought into relation.

The substantive concerns of urban planning are shot through with questions of social and spatial equity, and this might be seen as an overarching concern of planning: to moderate the balance of private and collective gains and losses. The self-interest of millions of individual households across the cities of the global south seeking to improve their lot is tolerable in this sense because it represents the pursuit of modest private gains by so many with so little. The search by clubs and even states for gains through urban planning is less tolerable since it often comes at the expense of larger majorities of the population, and it is to a consideration of such tragedies of the commons that I now turn.

The tragedy of the commons and other wicked problems

Urban planning problems are inherently 'wicked'. That is, they are ill-defined and rely on political or moral judgements rather than technical solutions (Rittel and Webber, 1973). Moreover, the moral or political judgements involved are complex and cannot neces-sarily guarantee the sorts of socio-spatial justice that we discuss above (Wildavsky, 1971). This has not prevented technical solutions continuing to be proffered and complex moral and political judge-ments cloaked for convenience in the garb of technical solutions. Is it morally right for planners or politicians to treat a wicked problem as if it were a tame one or to prematurely tame a wicked problem (Rittel and Webber, 1973: 161)? If individually we are often to be found making our best efforts to ignore them or wish them away, wicked problems are precisely the sort that urban planning – as a practice which seeks to balance individual and collective interests – cannot ignore.

The biggest question of all – that of the long-term prospects for our inhabitation on earth (see figure 4.1) – was famously and pithily captured in the notion of the 'tragedy of the commons' (Hardin, 1968). In the guise of climate change, the sustainable habitation of the planet is the best and biggest example of wicked urban planning problems, and one that refuses to go away. For Hardin, the tragedy of the commons is one in which 'freedom in a commons brings ruin to all', since 'the individual benefits as an individual from his ability to deny the truth even though society as a whole, of which he is a part, suffers' (Hardin, 1968: 1243). An alternative to the commons 'need not be perfectly just to be preferable ... Injustice is preferable to total ruin' (Hardin, 1968: 1247). The ever-present potential for a tragedy of the commons is a function of our existence on a planetary home with distinct resource and environmental limits.

The principle of the tragedy of the commons has been evident in numerous localized instances for some time, whether through the overgrazing of common land in the Middle Ages or what archaeologists speculate as the destruction of Easter Island by way of the deforestation required to move totems. As a problem recognized as one of more singular, universal import, it might be fair to say that it forced itself partially into our consciousness with Malthusian concerns of the late nineteenth century over the ability of the planet to sustain population growth, and with the same concern in the late 1960s. The concerns expressed in *The Limits to Growth* (Meadows et al., 1972) were quickly ignored in the face of economic recession in the global north and in the warm glow of new technologies of eco-modernization. However, the universality of our tragedy of the commons soon resurfaced in the guise of global climate change and questions of environmental sustainability in the 1990s. Climate change appears to provide an affirmation of the thought that 'we can never do nothing. That which we have done for thousands of years is also action' (Hardin, 1968: 1247). If heads of nation states and intergovernmental organizations have been slow to act, ambivalent or hostile to the evidence, much subnational urban planning policy and

Figure 4.1 Graffiti on a Melbourne road sign
Source: author

practice across the global north has been more successful in exerting a regulatory effect on resource use and the emission of pollutants.

The tragedy of the commons is not one that can be ignored. That is, arrangements for the rationing or regulation of commons of various sorts have become necessary. In comparison to the expansion of notions of justice implicated in urban planning to include now many different interest groups, we might argue that we have barely begun to address the extremely awkward, divisive and dangerous injustices that may come to be associated with the difficult political and moral choices associated with the tragedy of the commons.

History reveals public policy and urban planning's dealings with wicked problems. History warns not just of the inefficiency of reinventing the urban planning wheel but of the loss of urban planning knowledge and wisdom altogether. For 'education can counteract the natural tendency to do the wrong thing, but the inexorable succession of generations requires that the basis for this knowledge be constantly refreshed' (Hardin, 1968: 1244). To repeat my earlier warning: in an information age, the real danger is one of information overload and loss.

If the tragedy of the commons is a wicked problem of singular global proportions, it is true to say that every other lesser wicked problem is unique: 'There are no classes of wicked problems in the sense that principles of solution can be developed to fit all members of a class' (Rittel and Webber, 1973: 164). This once again reaffirms the need for the planning imagination to have a geographical sensibility; geographical context is important and a case-by-case approach is central to better urban planning interventions. It may be why we see the global tragedy of the commons that is upon us in the form of climate change and resource depletion being taken up at the local rather than national or international levels.

It is in this potential for local responses or experiments that Ostrom and colleagues question the inevitability of the tragedy of the commons, since 'the store of governance tools and ways to modify and combine them is far greater than often is recognized' (Dietz et al., 2003: 4). The tragedy may be a less bleak drama given the available evidence of successful management of the commons. The question of how to manage the commons is one of who plans: citizens, clubs or states. Ostrom and colleagues have championed decentralized self-organization in the form of altruistic clubs rather than big government or market (atomistic citizen action) solutions. Self-organized club solutions certainly appear workable for the management of small-scale (subnational) commons but may be less well suited to the grand challenges facing the global commons, for which there remain major difficulties in designing international, interstate institutions (Stern,

2007). Nevertheless, a way forward is even less likely without this sort of local urban planning experimentation.

Urban planning, externalities and land-value capture

The urban planning practised by individual households and clubs until the formation of nation states took place in a varied landscape of people's relationships, rights and obligations to land, but it began to be overlain by modern urban planning concerned with defining and apportioning land as private property. As such, the issues of compensation and betterment that flow from the private possession and uses of land and property have become fundamental to virtually every society in which urban planning interventions – in forms great or small, private or public, by citizens, clubs or states – take place.

Externalities

The geographical concentration of residents is perhaps the lowest common denominator of what we consider as a city. Cities in this simple form are hardly ever found since they immediately call forth the need for some form of collective organization. Collective organization is vital to provisioning the basic needs discussed in the previous chapter. It is also a response to 'externalities'.

In settlements, the actions of individual urban citizens can rarely be confined. Actions almost always 'spill over' to have effects that are external to the individuals initiating them. These 'externalities' can be positive or negative. The gathering together of individuals in geographic space tends to minimize costs of communication and transportation, which is a positive externality – a benefit shared by all rather than captured by a single individual. Equally, one person's or business's activities may pollute a neighbour, in terms of loss of visual amenity (access to sunlight) or the generation of excessive noise or airborne pollutants. The existence of cities provides general proof that, on balance, there must be positive externalities. If, on balance, the externalities were negative, there would presumably be no cities – population and economic activity would be widely dispersed.

Externalities can be classed as pecuniary or technological (Scitovsky, 1954). There are pecuniary externalities that affect the bottom line in measurable monetary terms. These are often apparent and are sought by some portion of the businesses that have typically congregated in quarters or districts of cities, since they are interdependencies that are traded. The term 'externalities' originates in descriptions of industrial agglomeration within particular districts or quarters of UK cities. Birmingham's jewellery quarter, now much denuded and the subject of urban regeneration policies, provides a classic example of the sorts of huddling together of businesses that continue to be visible in cities around the world. Its modern-day equivalent in cities of the global north might be the 'Tech City' software industry concentration in London and the high-technology industry in California's 'Silicon Valley'. Industry agglomerations – such as the batik industry in Laweyan, in Surakarta, Indonesia – have come and gone across cities of the global south, where others remain anchored in urban locations such as those found in the urban villages of Guangdong Province in China, garment manufacturing in Dhaka in Bangladesh, or the recycling industries in Dharavi in Mumbai.

However, some of the collective benefits offered by city living are hard to put a monetary value on. Much of the communication that is facilitated in cities is not instrumental – it has no immediate monetary gain. To the extent that cities are social places, they are places for the exchange of ideas, experiences, the development of knowledge and understanding of oneself, others, the city and the wider world. These externalities might be termed 'technological externalities'. A monetary value can be placed on some of these if knowledge can be codified and appropriated in the form of patents or licensing agreements, but otherwise knowledge remains embodied in, and distributed across, the myriad individuals found in the city. It is precisely these technological externalities that Jane Jacobs (1969) saw as the essence of the economic sustainability of the world's greatest cities. The most diverse urban economies – typically the largest or else best-connected smaller cities – are those that can draw on the collective inventiveness of their inhabitants in sustainable ways.

The longevity of cities themselves is one in which patterns of land and property ownership, externalities and socio-economic status play out over time. The overall balance sheet and distribution of externalities manifest themselves in the 'life cycle' of cities. Indeed, much of the orthodoxy of modern urban planning as it developed in the global north during the last century reflected the desire to decongest the city in light of the negative externalities of industrial growth: atmospheric pollution and traffic congestion. The widespread suburbanization of cities across the global north saw inner cities lose population and their land values decrease. Selective processes of gentrification which began in many of those same cities from the 1960s onwards have produced a return to the city, and with it a reversal of urban planning orthodoxy in favour of the virtues – the positive externalities – of density of development, architectural heritage and urban amenity. Here we can see how 'many of the externalities from land use have complex and inter-generational impacts' (OECD, 2017b: 29). It is the positive externalities that remain the draw of the city, much like the promise of London's streets for Dick Whittington.[1] It is the unanticipated or unavoidable negative externalities that imprison many a Dick Whittington. It is important that urban planners seek to recognize and redistribute the socially and spatially uneven benefits of positive externalities and incidence of negative externalities if cities are to be just.

Private property, externalities, betterment and compensation

The rendering of urban planning as the regulation and zoning of land use has been partly responsible for the distancing of urban planning from important questions of compensation and betterment (Blomley, 2017). Regardless of the aesthetics of the development produced, neither zoning nor a lack of it – as in Houston (Qian, 2010) – ensures equitable outcomes, in terms of the distribution of gains and losses in property values, without accompanying measures or plans to provide for compensation for adversely affected property, and for financial or land contributions to offset windfall gains. Different urban planning systems incorporate different solutions to compensating and taxing property owners, each of which is imperfect (Booth, 2012). For the greater good, on grounds of equity and social justice, the relationships between land and property values and urban planning is one that simply cannot be ignored, since 'land and the property built on it constitute by far the largest part of all global wealth' (OECD, 2017b: 28), as Piketty's (2014) *Capital in the Twenty-First Century* has underlined recently. And yet, lamentably, issues of 'land-value capture' (i.e. the capture of betterment) and compensation are ones that are frequently ignored or thought too difficult or too politically toxic to address, despite their major consequences for socio-spatial justice.

In 1271, in the Dutch city of Dordrecht, rulings decreed that the compensation price offered to landholders was not to exceed agricultural (current-use) value in a process of compulsory purchase. In 1558, in the Dutch city of Alkmaar, the completion of a new market entailed that a large number of surrounding land owners were to pay betterment towards its construction, as they were deemed to benefit from it in the future (Morris, 1994: 140). Somewhat later, the issues of betterment and compensation were the subject of vast philosophical debate, notably in the UK even in the pre-1914 era, and have remained largely unresolved today (Sutcliffe, 1981). Their consideration culminated during the Second World War as British society sought to draw lessons from the extensive government direction of resources that had been brought into being. The 'Uthwatt Report', produced in 1946 (Cullingworth, 1980), stands as perhaps the most comprehensive treatment of the issues of compensation and betterment ever made, and it had an influence on what urban planning was able to achieve in practice – in terms of the comprehensive redevelopment of towns and cities, the construction of social housing and the building of new and extended towns – during the half century after the war.

Betterment may be defined as that portion of an increase in land or property values not attributable to improvements made by a property owner. A host of old ('the unearned increment') and new ('land-value capture', 'developer contributions', 'proffers', 'planning gain') terms speak to this feature of the ownership of land and property. At first glance, betterment may seem an unlikely event generating modest gains to an individual land or property owner. This much might be implied in the term 'unearned *increment*'. Yet the gains from generalized increased land and property values can be substantial. Indeed, betterment can encompass several sources of land- and property-value increases that are 'unearned' by the efforts of the owner. Most prosaically, they can result from improvements made by tenants. They can result from the general growth of the local economy. However, they can amount to substantial windfall gains from the greater accessibility (and hence development potential) associated with urban planning decisions and interventions – namely publicly funded improvements in infrastructure, the zoning and rezoning of land for particular uses, or the granting of planning consent. The problem of valuing land is, to some extent, a counterfactual question of what would have happened in the absence of what is, in reality, a continual stream of urban planning decisions made by citizens, clubs and nation states. Geographical context rears its head, since the ability to tax betterment and to use it productively for the wider public interest is related to the economic health of the city concerned. Revenue from betterment taxes would be substantial in a buoyant local economy but may be non-existent in a declining or stagnant local economy. This itself can be one ingredient in divergence at the subnational level within a given national urban planning system (see chapter 6).

While the concept of betterment seems clear, it can be difficult to measure, let alone collect. In contrast, the concept of compensation is not only clear but easier to measure and hard to avoid paying (Crook, 2016). Yet the two cannot be separated when it comes to how betterment can best be leveraged in the service of urban planning for the just city. As we will see, both in principle and in practice, the participation of actors charged with acting in the public interest (by way of compulsory purchase and compensation of land owners) 'before the event' of development is the most effective way to leverage uplifts in land values, though it does not guarantee that these revenues would be used to plan for diverse and equitable cities. So it is that 'the mechanics of capturing betterment have proved troublesome, and ensuring that value is indeed diverted to serve the public interest has often been elusive' (Booth, 2012: 74). The key reason for this is the way in which our individual and collective relationships to land and property are 'socially constructed' – defined and redefined in law in ways that reflect broader societal values. By now a wide range of measures is used to tax betterment with greater or lesser effect across primarily wealthy nations (see, for example, OECD, 2017b). In principle, the most effective ways of capturing betterment cannot be separated from compensation paid, since they involve direct acquisition and banking of land by authorities acting in the public interest.

Interests in taxing betterment *directly* first surfaced in the UK against a backdrop of a revulsion against landed property (the feudal system of land holdings) in the nineteenth century (Booth, 2012: 77), and it remains the case that only through public ownership of land can equity be fully released (Fainstein, 2012). The 1909 Housing and Town Planning Act in the UK originally proposed to capture 100 per cent of land-value increases; this was reduced to 50 per cent and set alongside a countervailing provision for compensation in the face of political opposition. The generalized direction of resources by the state for wartime purposes lent legitimacy to the large-scale acquisition of land by the UK government during the 1940s and 1950s at or near to existing (rural) use values. This briefly created a dual market in land, as private purchases were at future development values. However, the social and political mood shifted decisively away from any serious attempt to link land-value capture to questions of socio-spatial justice after the 1960s, with debate surrounding the taxing of betterment and the paying of compensation increasingly centred on the efficiency of measures used. By the 1990s, the vacuum in national policy on betterment was being filled by local initiatives in the form of so-called section 106 agreements under the 1990 Town and Country Planning Act, which enabled local government to bargain for contributions from developers. Questions of betterment had been severed from national political debate and policy development to become ones of negotiation at the local level. Thus, the taxation of betterment in the UK has travelled a long way from its radical roots in concerns with socio-spatial injustice, to a concern with efficiency and a focus on a narrower suite of taxes that bring the UK closer to the systems in the US and Australia (with contributions or proffers and impact fees) and are noted less for their societal concerns for socio-spatial justice than for the rights of private land owners. The problem with *indirect* or 'after-the-event' measures is that they simply do not capture the scale of betterment needed to fund anything other than immediate site-related infrastructure or mitigation measures, and are, for example, totally inadequate as a means of funding the construction and maintenance of meaningful volumes of social housing. Ambrose (1994) notes how in Scandinavian countries the tradition has been one of state purchase and provision of land for development, while developers compete on price and quality of proposed schemes as opposed to being able to directly benefit from increases in land values. In Germany and the Netherlands, public sector land banking has been used much more extensively than in the UK, Australia and the US, where it has been private developers who have been accused of land banking, and hence effectively benefiting from uplifts in land values over time as well as the 'normal' profit margins on construction itself. In the UK, the merits of private versus public land banking, as these might impact housing supply and prices, have been ignored in a vacuum of national political leadership on the issue.

Land readjustment has attracted interest as one means of dividing up the spoils of betterment under conditions where fragmentation of land ownership might otherwise frustrate planning processes, where older or extant built environments can usefully be reworked, where major new infrastructure is needed and where public authorities or any individual land owner have insufficient resources to carry out development alone (Larsson, 1997). In Japan, land readjustment mechanisms have allowed private or quasi-private transit companies to use revenues from developing, selling and leasing residences and commercial premises to fund the initial infrastructure development. Each landowner contributes a portion (usually a third) of their initial holdings for provision of infrastructure, public space and a land reserve; the last of these is sold at the end of the process to pay the costs of planning, administration and construction (Sorensen, 2000: 219). While the mechanism can have some notable advantages in securing the development of infrastructure and a measure of transit-oriented development, it has promoted urban sprawl at a regional scale. An example here is the development now occurring along the TX line to Tsukuba, once an isolated science city and now increasingly an outer suburb of Tokyo (Miao, 2018a). Though difficult to negotiate among state, club and citizen actors, land readjustment in the form of land sharing can be one urban planning tool in cities across the global south, as a partial resolution to stalemates produced by successful resistance to forced evictions of slum dwellers, as indicated in cases in Thailand (Angel and Boonyabancha 1988).

In socialist states, such as China, where land is in communal or state ownership and is leased, the unearned increment in land values is fully captured for the public good if appropriately used by communes or the state:

> Perhaps the broadest and most comprehensive application of value capture is in China, where municipalities buy adjacent agricultural land from farmers at agricultural use prices, service it with infrastructure, and sell it to developers as urban land ... the difference in price between the land's urban value and its agricultural value accrues to the municipality. (Ingram and Hong, 2012: 16)

The issue is more one of possible corruption in the use of such windfalls to governments or adequate compensation to those whose lands have been annexed for formal urban development.

In the UK – where there is a mix of feudal land-ownership patterns, a liberal market in land transactions and a socialist system of rights to develop on land – the continued political need to construct the public interest is apparent. Successive governments have retreated from acquiring land at or near current use values and from using powers of compulsory purchase, and instead have sought to deal with a growing structural shortage of housing through developer contributions. In the US, state takings for public interest are so curtailed, and so subject to legal challenge in which courts routinely uphold the rights of private land owners, that a coordinative role for planning is increasingly reduced to suturing together a multitude of privately developed communities.

International dialogue in urban planning can and must reaffirm its centrality in helping to resolve such a fundamental issue in ways that balance private and social costs and benefits. It is all the more troubling, then, that although land-value capture mechanisms 'are common throughout the OECD, their fiscal impact is small and only small sums are raised through them' (OECD, 2017a: 10). Indeed, value-capture mechanisms found even across the wealthy OECD countries are considered 'rudimentary' (OECD, 2017b: 62). Urban planning will need to look to the lessons that can be learned from different parts of the world. These may be positive messages of what more can be done in the name of diverse and equitable urban planning, drawn from those contexts where strong popular political support and the bureaucratic will exists to tax betterment. Equally, the gross inequities in cities of the global south are at least partly a testament to the inability to extract developer contributions despite the free-riding or additional impacts upon inadequate infrastructures that numerous individual developments make. Moreover, urban planning will probably need to mobilize multiple means to capture land values effectively. Within the UK context, Crook, Henneberry and Whitehead (2016) argue that a two-pronged approach is needed which reflects different scales of urban development. For new, or major extensions to, towns, this might involve land banking by public sector authorities, with land bought at or near use value and serviced by new town development corporations, or the sorts of local government joint venture companies found in China. Remaining incremental infill development might be covered by obligations or impact fees of some sort.

Unintended consequences

Urban planning's achievements often come with unintended or unanticipated effects. In capitalist societies, planning can be viewed as an endless stream of piecemeal interventions (Scott and Roweis,

1977) whose unintended consequences are inevitable and increase in scale and scope over time.

Planning failures great and small

The unintended consequences of planning can come in numerous small-scale failures or large, discrete or 'great planning disasters' (Hall, 1980). An historical perspective contributes to evaluating planning's outcomes and probing the term 'disaster'. Great planning 'disasters' of the day – such as London's Millennium Dome, the Sydney Opera House, London's Canary Wharf and some of America's privately planned new towns (e.g. Reston) – have over time come to be regarded as great successes. Highly successful and widely emulated urban planning interventions such as the UK's new towns have barely been regarded as successes, at least in the home nation.

Unintended and unanticipated effects are present regardless of the particular urban planning actors involved. The inertial effects of the urban planning of modern nation states is a theme that Healey (2007: 287) draws attention to: 'the most risky element of spatial strategies arises if they become powerful shapers of future potentialities. Then, their inherent contradictions may close off emergent potentialities which later come to be seen as desirable.' The problems may be acute when the modernization sought by nation states is especially rapid, as in the transitions to and from socialism experienced in Russia and China. Here, 'a problem with large leaps of policy across many areas simultaneously, is that it may be impossible to reverse unanticipated, undesirable results' (Nolan, 1995: 57). However, rarely mentioned in the same hot breath of castigation reserved for the great planning disasters of states are the more numerous failures of individual citizens and clubs. 'The blunders of corporations or developers are often more difficult to detect early – and delay or prevent – than municipal or civic follies' (Rykwert, 2000: 230).

Self-fulfilling prophecies of planning for mobility

If mobility has been one enduring focus for the urban planning imagination, that imagination has often been restricted to viewing the demand for travel as a derived demand – the need to be somewhere else. Yet, by now, the limits of this view and its meaning for transport planning have been revealed in the way urban planning interventions catering to mobility routinely have 'rebound' effects. Improvements in technology and savings in time tend to generate *yet more* demand for travel: 'Thus, for every policy, be it technological or demand management, it would seem that complementary policies to "lock-in" any benefits are necessary' (Banister and Anable, 2009: 57–8).

Some of these unintended or unanticipated consequences of planning for mobility register themselves in the sorts of wildly inaccurate forecasts of travel trends that I discuss in the next chapter. Despite the sophisticated computational methods involved, the swathes of historic downtowns flattened to make way for highways in US and UK cities were often the product of wildly exaggerated estimates for car-based access to central cities.

Failures to plan with nature

The unintended side effects of modern state interventions have been systemic, though they are in some instances and in some parts of the world hard to disentangle from long-standing natural hazards that have attended urbanization. The disaster planning expertise that has been slowly accumulating in particular parts of the world – such as flooding, earthquake and bush-fire management in the Netherlands, Japan and Australia respectively – may yet become a more profound focus for the sorts of planning policy exchange I discuss in chapter 7. Here urban planning may be well suited to collective action problems that exist over a cycle of prevention, preparation, response and recovery from disasters, and where improvements in the recovery phase are still much needed (March et al., 2017).

Systemic urban planning failures are visible in the inability to heed the call to design with nature (McHarg, 1969). Considerable damage has already been done and continues to be done to the natural environs in which we have settled. Urban planning policies have failed to prevent flood plains being built on. Engineering projects bring water to otherwise water-scarce regions where large-scale urbanization might be unwise, not least in connection with bush fires. To be sure, these failures have called forth efforts to remake nature in cities such as Los Angeles (Gottlieb, 2007). Efforts to remake nature face the stiffer task of engendering a sense of stewardship of the land when some of those would-be stewards, such as farmers or estate owners, are increasingly promotors of development at the urban fringe.

In the case of bush fires, for example, the failures of modern urban planning may signal a need to relearn the ancient knowledge of land management held by indigenous populations.[2] By the same token, they provide a motivation to incorporate those same indigenous populations more fully into planning systems which have remained impermeable to them, their claims to land and the knowledge that adheres to it.

Inabilities to control urban sprawl

Processes of suburbanization might be defined as 'a combination of non-central population and economic growth with urban spatial expansions' (Ekers et al., 2012: 407), and urban planning, more than ever before, will have to be about *suburban* planning and have an imagination attuned to the suburb in all its forms. The expanses of new residential suburbia that now make up the major part of our cities are something that urban planning has sought, with mixed results, to re-enchant (Knox, 2008). The enthusiasm engendered by the New Urbanism movement in the US is associated with strong interest among the architecture and planning professions in retro-fitting or repairing sprawl: 'the redevelopment of sprawl into more urban, more connected, more sustainable places is the big project for this century' (Dunham-Jones and Williamson, 2009: v).

Much of this interest is focused at the building or site-specific scale and fails to develop the sort of political and planning imagination necessary to address the sheer size of the potential urban planning challenge that is unfolding (Phelps, 2015). Suburbanization as a process draws upon desires that urban planners would do well not to ignore but to understand and better plan for (Bruegmann, 2005). However, the properties and potentials of suburban formats of development are complex and remain poorly understood (Kolb, 2008). Suburbs are often far more diverse in their populations and land uses than is commonly appreciated and almost certainly present more complex mixes of the sorts of urban planning actors we are concerned with here: citizens, clubs and states. New suburbs have an international variety that represents something of a frontier for the development of new urban planning thought, methods, practice and, ultimately, wisdom.[3] The implications of this variety of global suburbanization are coming to light at the same time as we continue to ape past suburban environments. The single case of California's

'Silicon Valley' outside San Francisco has been invoked in urban planning visions from Tokyo Metropolitan Government's 'TAMA Silicon Valley' of the 1970s, to Malaysia's 'Multimedia Super Corridor' outside Kuala Lumpur (Bunnell, 2004) planned in the 1990s, to the present of Hangzhou's 'Future Sci-Tech City Corridor' in China (Miao et al., 2019).

Powerful distinctions in urban planning thinking and policy (such as urban and rural) have failed to stop in-between patterns of urbanization (Phelps, 2004; 2017; Sieverts, 2003). These in-between spaces suffer from a deficit of planning and remain little understood and largely ignored in the urban planning imagination. If anything, the various in-between spaces that we can observe today – urban interstices, suburbs and metropolitan-scale 'rurbanity' – present the most institutionally complex landscapes for urban planning to operate on, since they are those in which there is often the poorest mutual understanding and articulation among the different citizen, club and state actors involved. Urban informality from global south to global north is distinct in its emergence within the interstices of cities. Across the global south, 'informality seeks the vacuums of urban space – the leftover, absent spaces and buffer zones; the unsafe, unusual or unstable' (Dovey and King, 2011: 26), with these interstitial forms specified as not only 'urban waterfronts and escarpments, but also the interstitial easements lining transport infrastructure and the deeper urban spaces behind formal street walls' (Dovey and King, 2011: 11).

Across the formalized and strongly enforced urban planning systems of the global north, the absence of the planning imagination registers in a different way and at a different scale. Here, the planning framework of extant, largely built-out, urban jurisdictions and expanding environmental designations has a clarity that the spaces where new urban development, population and employment must go rarely have, as in the case of London and the southeast of England in the UK (figure 4.2). And yet, since urban planning is essentially decentralized to numerous local governments, there is little agreement or purposive planning for these spaces. Recent UK government policy and incentives to fashion joint working among local governments to deal with 'beyond boundary' issues, including in the realm of urban planning (in the form of multi-area agreements and combined authorities), do not necessarily speak to these spaces.

'Between the obvious poles of extreme concentration and extreme dispersion, there is a vast range of form that is often difficult to

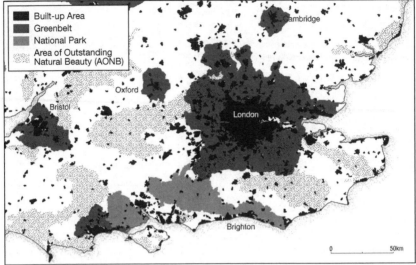

Figure 4.2 Pathways of urban development in the southeast of England
Source: WikiCommons and Open Street Maps

characterize as definitely one thing or another' (De Jong, 2013: 15). Urban planning must develop the imagination and policy vocabulary to better address these interplaces (Phelps, 2017). Patterns of urbanization in Germany have been the inspiration for Sieverts' (2003) use of the term *Zwischenstadt*, or 'in-between' territory, to describe a new urban landscape that is visible around the world but is barely recognized and poorly planned for. The fundamentally mixed or 'rurban' settlement pattern found across East Asia has been termed *desakota* by McGee (1991) in a joining of two Indonesian words: *desa* (village) and *kota* (city). The *desakota* pattern has rarely been thought desirable in urban planning terms but has been a reality posing distinct planning challenges for decades in countries such as Japan, where it manifests in the fine grain of urban agriculture found in Tokyo, and in vast stretches of ribbon development in many parts of Indonesia that planning is often ill-equipped to regulate.

Participation and the lack of it

Urban planning, as we have seen, is involved with questions of the public interest, however hard that may be to identify as something that exists factually, let alone as a criterion or value that urban planning can respond to. Yet it is some of the difficulties of defining the public interest with respect to urban planning that place it, as Friedmann (1987) suggested, squarely in and constitutive of the public domain. One of the key dilemmas of urban planning thought and practice is how to generate participation or dialogue as a reflection of our being in the world; participation in decisions of city building *as if* the public interest existed as something exceeding the self or particular and competing interests that patently do exist. Dialogue, participation or communication are central to urban planning as an activity that reflects our being in the world and the inevitability of our encounters with others, and it is a property needed regardless of the actors involved, the era or moment in question, the scale of place or non-territorial relation, or the substantive concern involved. A major contribution of urban planning to urban society is to ensure there are enough eyes, ears and voices present to properly scrutinize, oppose, and propose alternatives to plans and development proposals (Tiesdell and Allmendinger, 2005).

Questions of who plans, by whom, for whom, and who is included and excluded have become the equivalent of a search for a Holy Grail within planning, but participation presents as many problems in the planning process as its absence. Famously, as we saw in the preceding chapter, it can – as in parts of southeast England – result in significant NIMBY ('not in my back yard') and BANANA ('build absolutely nothing anywhere near anyone') sentiments that inhere within coveted planning policies (such as green belts), but which often confound provision for basic human needs (such as the right to shelter).

If local populations can prevent substantive issues being addressed like this, then in other instances urban planning may be able to leverage the positive demands of grassroots movements, as in Madrid (Castells, 1983; Phelps et al., 2006), Los Angeles (Gottlieb, 2007) and the *machizukuri* neighbourhood planning model that emerged in Japan (Evans, 2002). Indeed, the urban informality that characterizes the global south is an economic and social resource (Roy, 2005; Holston, 2009). Here, the likes of participatory budgeting have recently provided a window on the value of public participation in planning exercises, as I go on to discuss in chapter 5. 'Contemporary spatial planning approaches combine a curious mix of the more participative alongside the more deregulatory' (OECD, 2017a: 52), and stem in no short measure from the different actors involved.

Citizens and clubs such as defined interest groups typically do not organize or advocate for something, but rather against something (Rykwert, 2000). The increased participation by citizens and clubs has called forth greater participation from other clubs in the form of vexatious lawsuits anticipating or taken against the possible effects of public participation (Rykwert, 2000: 230).

Public participation has only been an element of planning for part of modern planning's existence, and there are difficulties associated with attempts to incorporate more participatory, rather than representative governance modes into planning processes. The importance of participation is seen across the different substantive concerns of urban planning. While the environment has been a concern within urban planning and often enshrined in strong urban planning policies and regulations, it is also the case that it is associated with self-deceits, as we saw in chapter 3. But the natural environment and our relationship to it have rarely figured as profoundly in participation in urban planning processes as this suggests. As Beatley (2016: 3) argues, 'human beings need contact with nature and the natural environment. They need it to be healthy, happy, and productive and to lead meaningful lives. Nature is not optional, but an absolutely essential quality of modern urban life.' As he goes on to observe, 'for nature's presence to be meaningful, residents must be aware of it, care about it, and engage with it in some way' (Beatley, 2016: 33–4).

Conclusion

It has been argued that urban planning has become more integrative of the diversity of different needs and wants present in the city and, as a result, developed more into a 'people service' (OECD 2017b). However, 'these twenty-first century roles of planning are not sufficiently recognized by either politicians or the public who, for the most part, see planning in its historical light ... It is these integration and democratic roles of planning that ensure planners' survival in the twenty-first century. These roles have given planning an implicit legitimacy' (OECD, 2017b: 59). A reinvigorated urban planning imagination will need to draw not only on past wisdom as it has emerged from cities of the global north, but, of necessity, on the future possibilities inhering in the participatory, co-productive and insurrectionary agendas emanating from the global south – of which more in the next chapter.

The wisdom of urban planning is that of the fundamental interconnectedness of our actions. It is the wisdom of what Heidegger (2010) described as our sense of being in the world, and what has more recently been described as a relational sense of place and community. The tragedy of the commons and other wicked problems reveal this, and questions of equity and justice in our cities speak to this. Dry and sometimes legalistic issues of compensation and betterment reveal the materiality of it. The unintended and unanticipated effects of urban planning itself show the complexity of the cities we build and inhabit. In our participation in the planning of our urban communities, we are continually confronted by the worst and best of our nature as we seek to resolve – however temporarily – fundamentally wicked problems. Urban planning's wisdom is that we must ensure that we continue to plan our cities with our better nature to the fore.

5 Methods: what are the means of planning?

Introduction

The foundational mantra of modern town planning – 'survey, diagnosis, plan' – has been attributed to Patrick Geddes as a founding figure of the discipline (Meller, 2005); 'monitor' is a later addition to this trilogy. It is important to note that the survey was a 'survey before plan' when compared to the evidence bases that are often assembled concurrently with plan making today (Batey, 2018). It is hard to escape the context in which this mantra found favour, coming as it did on the back of Charles Booth's study of London and Seebohm Rowntree's study of York, which laid bare the squalid conditions of urban life in the UK by the late 1800s. The status of 'survey, diagnosis, plan, monitor' was elevated rapidly, being promoted in 1909 by Raymond Unwin (who was influenced strongly by Geddes) and developed by 1916 by Patrick Abercrombie – most famous for the Greater London Plan (with its associated green belt) of 1947.

While Geddes' notion of survey was broad and inclusive, much has changed since this initial codification of modern urban planning. The geohistorical sensibilities implied in his survey had given way by the 1950s and 1960s to a less place-based focus on quantitative forecasting and analytical techniques. That is, the integrative role of urban planning had largely been lost by the 1960s (Batey, 2018). Calls for greater participation emerged from at least that decade onwards in the global north in the form of NIMBY ('not in my back yard') pressure-group reactions to plans and individual development proposals, and in the last several decades in radical forms such as grassroots 'right to the city' and alternative environmental and economy movements. These in turn have led to calls for more participatory forms of urban planning – variously labelled communicative, deliberative or collaborative – to be developed within representative democratic structures. At the same time, in the global south, genuinely participatory forms of urban planning and the co-production of built environments have been pioneered within representative democratic structures, often under conditions of austerity (Mitlin, 2008). Informality across the global south strongly implies the direct action of citizen-conceived, self-help planning of rapidly assembled neighbourhoods, towns and cities.

Just as I suggested that the who of urban planning is varied and includes states, clubs and citizens, so too are different methods of urban planning varied, each with its own distinctive strengths and weaknesses. The history of the methods associated with urban planning is not a linear one in which successive methods are replaced. There are important slippages or lags in the methods typically used in different parts of the world, with urban planning in rapidly urban-izing countries like China or India marked by aspirations to the methods by which global north urban forms have been arrived at. At least some of this is a product of the stage of urbanization reached across countries of the global south, since, for all their faults, such techniques are appropriate to the build-out of cities. Moreover, in many national and local contexts where urbanization rates have slowed, these different methods coexist and to an extent compete in terms of their claims to validity and legitimacy.

These competing claims and strengths of urban planning methods pose important questions regarding the future of the urban planning imagination – questions of whose hands it should be in and to what ends it should be turned. Taken together, they represent a reservoir of artistic and scientific approaches to tackle wicked problems that is internationally dispersed across citizens, clubs and states. To the extent that productive mixes of the sorts of methods typically deployed by different actors have not yet been exhausted, the variety of methods explored here offers hope for our urban present and future.

City as organism: the art and science of urban planning

Patrick Geddes 'instinctively appreciated the organic complexity of cities, liberally using analogies between urban and biological systems in both directions' (Batty and Marshall, 2012: 24). He was particularly keen that town planning should be both a theoretical and a practical affair, arguing that 'the methods advocated for the systematic study of cities ... be not merely the product of the study, but rather be those which may be acquired in course of local observation and practical effort' (Geddes, 1904: 104). The method he advocated as the study of towns and cities in advance of practice was broad and inclusive. It drew on the immersion in, and close reading of, extant literature on the topography and history of towns and cities and surveys of secondary or primary data. He described urban planning as an 'experimental endeavour'; as 'the art of enhancing the life of the city' (Geddes, 1904: 111). Urban planning was part art and part science, with each informing the other: 'our everyday experiences and commensurate interpretations gradually become more systematic, that is, begin to assume a scientific character; where our activities, in becoming more orderly and comprehensive, simultaneously approximate to art' (Geddes, 1904: 104).

Geddes' approach has been credited with breathing vision 'into an activity which, in the hands of architects, engineers and surveyors ... threatened to be concerned exclusively with a simple ordering of the physical environment' (Meller, 2005: 181). In viewing city-regional development as an organic whole, he recognized the 'betweenness of place' (Entrikin, 1991), bringing not only town planning theory and practice close together, but also academic geography and sociology into dialogue with planning in ways that have not been common recently (Phelps and Tewdwr-Jones, 2008).

Geddes' organic and evolutionary perspectives on cities and urban planning were not without some important tensions, since 'town planning perhaps rested on the town being simple enough to be able to plan, but complex enough to need planning' (Batty and Marshall, 2012: 25), implying the city as organically emergent and artificially planned at one and the same time. Nevertheless, the organic conception of cities and their regions has endured and continues to offer much as a planning wisdom of its own.

The survey element might be thought to lend itself to immediate urban problems. However, Geddes' viewpoint was always more medium to long term rather than short term (Meller, 2005: 56–7). This as much as anything saw the survey aspect of this foundational method become shorn of its qualitative and interpretative aspects within planning practice. The plans being produced in the UK after the Second World War lacked detail and consistency and were uninspiring. The fact that planners were at that point overwhelmed with guidance on plan making (Batey, 2018: 53) may be instructive, as too much stricture has served to blunt the creativity of local government planners. The 'survey method' became more narrowly based in the 'figure work' associated with the likes of the population projections justifying housing and employment land releases, or projections of travel patterns used to justify road construction. As Meller (2005: 324) notes, 'even while planning students were still being trained in Geddesian techniques ... the most essential element of Geddes' message, the critical relationship between social processes and spatial form, was ignored in relation to the existing environment'.

The organic metaphor featured strongly in at least one important urban planning tradition in the US: the Regional Planning Association promoted from the 1930s by Lewis Mumford, among others, and featuring in major plans for the Tennessee Valley Authority (TVA) and the City of New York and its environs. In practical terms, representing the organic whole of city regions in the built environment disciplines of urban planning and design continues to some extent in the technique of map 'overlays', by which different regions or scales of environmental and social and economic organization can be superimposed. The possibilities for urban planning to shape urban development in organic terms endures. As Thompson-Fawcett (1998: 168), commenting on Leon Krier's invocation of organic metaphors, notes, 'with an irreducible metaphor such as the organic metaphor, there are always fresh connections, creative extensions and newer turns that can be made'. It should be little surprise that organicism is therefore malleable as an analytical perspective and guide to planning practice. Kostof (1991: 75), for example, notes how 'it is a strange twist that "organic" patterns of antiquity and the Middle Ages were intimate frames where the rich and poor were woven together. The organic of the modern suburb is exclusive: this is a private world peopled by one's own kind.'

City as system: the predicting, providing and nudging of the state

Much of the sensitivity to the uniqueness of place found in the 'art' of urban planning had by the 1960s been shorn away in favour of exclusively quantitative, 'scientific' approaches to surveying and forecasting of trends and future demands in terms of key demographic and economic facts. The territoriality of the nation state ensures that its logic is inextricably linked with the techniques of surveying, codifying and quantifying those aspects of Cartesian space and accurately measuring what lies therein (Blomley, 2014; Elden, 2005). The state sees in a way which places a premium on these sorts of methods (Scott, 2000), to such an extent that modern urban planning might be seen as part of the tyranny exercised by all nation states at one time or another (Yiftachel, 1998).

Forecasting and a battery of quantitative analytical techniques dovetailed with ideas of the planning of cities as urban systems, as championed in Brian McLoughlin's (1969) *Urban Planning: A Systems Approach*. Those systems were – by comparison with what we understand today – treated as simple, closed and predictable; amenable to control in a fashion advocated in cybernetics as it was developed earlier in the disciplines of engineering and management (Batty and Marshall, 2012). For Batty and Marshall (2012: 28) 'the systems approach to planning, when it finally came, represented one of the last gasps of a movement established in the late nineteenth century. It looked increasingly irrelevant in practice in a world that was rapidly changing.' This may be true of the forward planning and development control practised at local government level. However, some of this approach necessarily retains a salience in the regional- and national-scale settlement and infrastructure planning vital to setting the context for successful local place making. Systems theory combined powerfully in the minds of some planning theorists with notions of a generalized 'strategic choice' method of decision making, in ways that meant that creativity regarding possible futures became neglected in conceptions of the plan-making process.

The methods of planning the city as a system that had emerged from the 1960s to the 1980s are familiar to us as 'predict and provide' or 'trend planning' (Pickvance, 1982). Some of the vagaries of forecasting quickly became apparent, in that forward planning on horizons of up to twenty years involving new allocation or zoning of land for residential and employment uses could be undermined not only by the normal fluctuations of business cycles but also by major global recessions. The preparation of these sorts of plans was so slow that the plans were already out of date by the time they were formally adopted. Forecasts of significant increases in population and economic activity in the south Hampshire area made at the end of the 1960s boom in the UK were exposed as grossly overstated by the time the plan was released seven years later (Phelps, 2012a). The redundancy of expensively produced plans at this time could be truly spectacular. In a similar period, the population of Kinshasa in the Congo increased from 700,000 in 1965 to 2,500,000 in 1977 as a new master plan was being produced (Beeckmans and Lagae, 2015). Plans of this nature and of this era were often characterized by a lack of the synthesis sought by Geddes, being instead a series of separate chapter surveys of conditions and forecasts (Batey, 2018: 56)

Despite their best efforts, state planners have often made spectacular failures of forecasting due to inappropriate assumptions and poor knowledge of the starting points of the trends they were trying to project. One striking example comes from road planning for London undertaken in the 1960s, where demand for an inner London motorway 'box' (begun in the 1960s but only partially completed in the face of protests) was massively overestimated, but the demand for an outer London orbital motorway (completed in 1986) was massively underestimated. The field of transport planning provides examples of the more fundamental, philosophical limits of trend planning. Here it became apparent that 'predict and provide' proved a self-fulfilling prophecy, where more road capacity led inexorably to more traffic and not to the resolution of congestion. In the UK, the predict-and-provide philosophy to road planning has only just been overturned, but more as a result of national budget constraints than anything else.

Smart cities: the city as a mine of data

Something of the systems approach lives on today as one concerned with 'complex' and radically 'open' systems, and with the desire for authorities to be smart in exploiting an explosion of real-time or big data on all aspects of city life. In this reincarnation of the systems approach to urban planning, vast amounts of real-time data allow tracking, analysis and some measure of prediction of the self-organizing city (Batty, 2018). Rather than single, once-and-for-all, 'silver bullet' planning solutions to urban problems, urban planning might be characterized as incremental but rapid adjustment to and modification of emergent trends – what we might term 'nudge planning'. Here there are parallels with 'nudge economics' (Thaler and Sunstein, 2008), in which policy seeks to understand the evolution of aggregate behaviour and nudge that behaviour at the margin. However, as with nudge economics (Hausman and Welch, 2010), important questions remain about the novelty, democratic credentials and ultimately the intelligence of this new systems planning.

As Batty and Marshall (2012: 34) argue, the city is not an organism but an ecosystem, where an ecosystem is 'composed of co-evolving sub-components ... indefinite in extent ... never in equilibrium'. Urban systems planning, rendered as 'complex systems', generates important insights relevant to interpreting and reflecting on the history of urbanization and urban planning. For example, the sorts of unchecked and exponential growth apparent in complex open systems mean that limits can be breached, resulting in significant historical discontinuities in trends. The complex-systems perspective underlines the thought that initial conditions can be difficult to identify. This 'new' systems approach, again, is popular largely at the level of analysis rather than practice and at the urban-regional rather than the neighbourhood scale. While the complexity-theory context for the communicative, collaborative and deliberative planning ideas and methods I discuss below is recognized by parties on both sides (e.g. Batty and Marshall, 2012; Innes and Booher, 2010), an enormous gulf exists between them.

The interest in smart cities is in making use of the massive amounts of real-time data that are available across many aspects of life in and functioning of the infrastructures of cities. Smart-city policies and strategies have by now been developed with distinctly different flavours around the world. In Singapore, the smart-city notion has been allied to long-standing top-down and expert-led urban planning, and designed to maintain the city state's lead in the smart and sustainable urban solutions field as an exportable service (Miao, 2018b). In contrast, in the administrative complexity and liberal economy context of New York City, the traditional economic advantages of urban agglomeration as an incubator of enterprise and innovation are hoped for from numerous state, club and citizen innovations. In Helsinki, the desire is to open the city as a data platform in which information is freely available for use in all software applications, whether developed by interested citizens or by the private companies that city authorities hope will flourish in the wake of the contraction of national telecom champion Nokia. In Barcelona, smart-city policies have lurched from an engagement with the major tech corporations that have pushed hard to enter lucrative smart-city hardware and software markets across the globe, to being much more rooted in the democratic traditions of the grassroots political movements that emerged under the Franco dictatorship (1936–75). Here, the city is now viewed as a data platform providing for the flourishing of citizen groups and mass participation in decision making for the city. Strategies extend to groups of cities, but significant problems remain with the compatibility and sharing of data within individual cities, let alone between cities, where newly formed alliances have struggled to become platforms for national- or regional-scale settlement planning, as with the case of the Scottish Cities Alliance (Miao and Maclennan, 2019).

In all of this rush to 'smartness' among analysts and practitioners, there is something profoundly antithetical to the urban planning imagination. The urban planning promised by smart technologies and associated municipal strategies might be a good example of the wisdom of incrementalism celebrated by Lindblom (1959). However, it may amount to the more efficient delivery of the present rather than purposeful planning for a better future. One of the reasons for this is that while most human encounters can be read as interactions, 'the only thoroughly uninteresting qualities of interactions, which are generally not worth valuing at all, are their number and frequency' (Stretton, 1975: 285). These are precisely those qualities that come to the fore in smart-city thinking. At its worst, urban planning may be implicated in the sorts of manipulation by algorithm that have become apparent with the rise of advertising and provisioning via the internet. It may amount to nothing more than a civic surrender to software code (Kitchin and Dodge, 2011) to place alongside an earlier surrender to zoning code (Talen, 2011). The absence of the urban planning imagination to guide and reform the coding of our cities should be truly a cause for concern. For some there is cause for optimism in the thought that we may be in a historical moment in which new zoning codes are being enacted that undo the worst excesses of land-use separation promoted previously (Ben-Joseph, 2012).

The rendering of urban planning as the control of systems was quickly exposed not only by some of the dramatic failures of its own methods but also by a change in the zeitgeist. The smart-cities incarnation of systems thinking will probably be similarly exposed in the future.

City as scenario: clubthink

'If planning is so ineffective, why do we do it? And why is it so important that we continue to do it? ... At its best, planning becomes a form of anticipatory ... thinking' (Farson, 1996: 125). These sentiments are echoed by Hoch (2019: 13) when speaking of plan making as a means of anticipating, coping with and preparing for complexity. Much of this sense of urban planning as a form of anticipatory thinking is found in the method of scenario planning. Although intriguing, the assertion that 'scenario planning is what we do as human beings all the time' (Lindgren and Bandhold, 2009: 1) probably does not ring quite true for many citizens. In the global south, extreme precarity continues to confound much anticipatory thinking for much of the time for citizens in their daily lives. While there are citizen aspirations that it is important to recognize in academic analysis and cater to in urban planning practice (Ferguson, 2006), these are unlikely to find a place in the sorts of scenario planning that are most closely associated with clubs. Equally, it could be argued that the pervasive role of the state in global north urban planning means that the extreme ordering of the built environment obviates much of the need for, or precludes, citizen scenarios.

Scenarios are 'creative, grounded explanations of change, developed through concerted interaction with constituencies as sources of knowledge and legitimation' (Hopkins and Zapata, 2007: 14). A scenario is not a forecast, and it is not a vision. Scenarios are 'stories about how the world changes' (Hopkins and Zapata, 2007: 9). Following Lindgren and Bandhold's (2009) primarily business-corporation-centred discussion, it could be argued that scenarios have their own logics. Visions are about desired end states and mobilize the power of seduction – something I return to in chapter 7. Forecasts are about extrapolating present and past trends and are usually undertaken by experts and rely on administrative fiat, as we have discussed immediately above. Scenarios arguably rest on a communicative logic, involve reciprocity and mutual learning, and mobilize relational power (the power to persuade and enlist others into a project).

Futures research had its origins in the Cold-War-related consulting initiated by Herman Kahn and the Rand Corporation in the 1950s, and later, and with extension to civilian applications, with the Hudson Institute in the 1960s. It had filtered through into private companies like Shell by the 1960s and 1970s and to governments by the 1990s (Lindgren and Bandhold, 2009). It has since shed some of its technocratic associations, to be deployed more widely by management consultants, governments, NGOs and academic researchers, including in the urban planning field, for specific substantive issues such as infrastructure planning and land-use futures.

The French regional economic planning tradition exerted an influence within the Europeanization of urban planning through the European Spatial Development Plan (ESDP) process precisely because of the power of scenarios concerning the development of EU space economy (Faludi, 2015). Spatial scenarios have been extremely powerful here in generating a measure of cohesion across a vast territory that might otherwise be subject to divergent economic trajectories. The series of metaphors invoked have become progressively larger and more inclusive, from the 'golden triangle' observed by the end of the 1960s, to the 'blue banana' and 'pentagon' of the 1990s, to the 'bunch of grapes' (Kunzmann, 1996). Here urban planning is left to provide the hope of cohesion in the face of the reality of uneven development. The work here continues through ESPON, which provides a database and reference base for urban planners across Europe and deploys scenarios as part of its advocacy role within the EU (Faludi, 2015). Scenarios have been actively used by the UK government, given the accepted complexity of actors and processes that influence land use. The Foresight Land Use Futures reports have engaged with scenarios in which climate change may prompt major redistributions of population and jobs in the UK away from London and the southeast of England. Indeed, climate change and the rise of the sustainable development agenda have generated an interest in scenario-based planning methods partly as a result of their perceived value in generating consensus (Wilson, 2009).

The difficulties that arise in generating good scenarios speak once again to urban planning as part art and part science.[1] Scenarios are stronger in certain contexts and under certain conditions than others. They are useful in planning for extreme events, since scenarios can expand thinking, especially with regard to the readiness and preparedness of communities for change. They are thought to uncover near-inevitable events in the future. Thus, some urban planning problems are essentially predetermined in ways that are predictable and may to an extent be manipulated. Perhaps the most important of these is captured in the pithy thought that 'demography is destiny'. The value of scenario planning here may seem trite until one realizes that major inflection points in trends that seem obvious with hindsight – such as Japan's ageing and declining population (Ohashi and Phelps, 2020) – have often been overlooked at the time.

The McKinsey summary suggests that methods of scenario building can protect against 'groupthink' or 'clubthink' and may be better at allowing people to challenge conventional wisdom, including established urban planning methods themselves. However, the methods associated with scenario building also have some potential weaknesses.[2] Interestingly, these include the sorts of paralysis that afflict seemingly every type of urban planning method. The sheer complexity of generating and selecting valid scenarios can lead to inaction. Scenario planning by central governments has tended to live a life of its own as a purely intellectual exercise (Lindgren and Bandhold, 2009: 28). As with the UK land-use futures exercises, scenarios can become divorced not only from the daily stream of development control decisions but also from the forward planning taking place at local- or regional-government tiers.

The rise, fall and rise again of scenario planning is an example of the non-linear history of urban planning and its repertoire of methods. A practice that came out of the Cold-War context and was abandoned for a time has become quite popular again among clubs and states. It would seem that once again an unashamed interest in futures can be a part of the urban planning imagination.

Cities and citizens: participation in planning

The limits of the expert-driven, top-down nature of urban planning as it was being practised by the 1960s in the likes of the UK and US were noted by Jacobs (1961) and Davidoff (1965), but they were apparent to and appreciated by urban planners as a lack of public participation in planning. Statutory planners and academics had themselves been aware of the largely tokenistic nature of public participation in urban planning processes (Arnstein, 1969) as it was being practised in the global north by the 1960s. Major government inquiries, such as the Skeffington Committee (1969) in the UK, calling for greater public participation, emerged in advance of sterner academic critiques (Simmie, 1974). Initially at least, academic critiques of systems planning offered little in the way of solutions to the issue of citizen participation in urban planning decisions. Moreover, the role taken by and envisaged for citizens participating in urban planning is complex and varied and might be thought to have several rather different forms with different implications, ranging from 'orthodox' methods of incorporating citizens' views into planning to more radical or activist experiments that promise outright opposition to state and club forms of planning.

Public participation

It might be argued that public participation in the urban planning processes associated with the representative democracy of national and local governments remains something of a holy grail despite the overwhelming evidence of its failure over many years in numerous guises. It remains difficult to incorporate the rich, lay, place-based knowledge embodied in numerous citizens into urban plan making and deliberations on individual developments, despite the importance of doing so (Hayden, 1997). The Skeffington Committee (1969) in the UK sought to increase public participation in plan making, setting out processes for the incorporation of public views. Today, such formal participatory processes are one of the few commonalities found across European urban planning systems (ESPON, 2018: 26). Yet, as the same review of European practices makes plain, most of the innovation with regard to promoting public participation occurred before 2000, with subsequent reforms emphasizing simplification, the creation of greater certainty, and reductions in administration burdens, to the point where public participation rarely exceeded what could be described as 'partial' in most EU countries (ESPON, 2018: 39). Formal planning processes continue to struggle to attract large-scale citizen interest, and Davidoff's (1965: 334) lament that 'there is something very shameful' about the necessity of organized citizen participation continues to ring true.

Strategic spatial plans that set out positive proposals for the long term and at scales which exceed the daily lives of people – cities, metropolitan areas or regions – struggle to command much interest among citizens. Surveys of samples of populations have been used to garner input to and feedback on draft local plans. Draft urban land-use plans are typically placed on exhibition or deposit by local governments, with comments invited and incorporated into subsequent iterations. Much consultation of the public was previously done in the physical form of mailed surveys and written submissions but is now facilitated online via website portals.

In liberal market planning systems, digital copies of plans are freely available, but it may be hard to locate a physical copy of the plan or a city model on public display. That the display of the plan and planning deliberations have both become more localized and site- or project-specific is in many ways all to the good, but something of the integrative or synoptic power of planning and the plan may have been lost. The situation contrasts with that in Singapore, where the Urban Regeneration Authority proudly retells the history of urban planning there to experts and tourists alike. Enormous, purpose-built 'cathedrals to urban planning' are to be found in cities across China. Cities that appear to suffer from deficits of urban planning boast planning galleries and aspire to be planned, as with Bandung in Indonesia (see figure 5.1).

Although all power is in the hands of the people according to the constitution of the People's Republic of China, it is of course well known that urban planning power has overwhelmingly been wielded

Figure 5.1 Inside the planning gallery in Bandung, Indonesia
Source: author

by local party members and government officials in conjunction with independent city and provincial planning institutes, individual university-based urban planning professors and, increasingly, international planning consultancies. Despite the emergence of public participation within urban planning across China by the 1990s, it remains at Arnstein's (1969) level of tokenism, according to one recent review (Institute for Sustainable Communities, 2015: 17). Nevertheless, as the range of actors incorporated into top-down planning has grown, as citizens have become more familiar with protesting redevelopment schemes, as neighbourhood committees have been constituted to replace *danwei* company management as a means of fostering loyalty to the Chinese Communist Party (CCP) at the local level, and as new grassroots participatory organizations have begun to emerge, so too has a greater measure of public participation, including experiments with crowdsourcing and NGO-led participatory planning exercises. Some of the examples, as in Nanshing District, Yining City, in Xinjian Autonomous Region, reflect the difficulties of conventional top-down planning methods. Others, such as the threatened redevelopment of the coastal tourism-centred urban village of Jiaochangwei, Shenzhen, have mobilized the most up-to-date crowdsourcing thinking and technologies (such as WeChat) to help collect information (Institute for Sustainable Communities, 2015).

At the same time as formal consultation processes often struggle to attract the interest of the public as a part of representative politics, interest groups have been particularly innovative in influencing urban planning deliberations indirectly. A host of participatory channels or points of pressure, often occurring through social media and direct action, now bear on the urban planning imagination as sources of input into decision making. Statutory urban planners are ever more subject to the competing evidence bases produced by such interests. Local groups are often extremely well-educated, resourced and effective adversaries of individual development proposals, to the point where the acronyms NIMBY ('not in my back yard'), BANANA ('build absolutely nothing anywhere near anyone') and LULU ('locally unwanted land use') are now familiar. Some of the same pressure-group politics has become increasingly apparent at examinations in public of local plans and inquiries into major road or airport building schemes – though the restrictive remit of these can see participation spill over into civil disobedience.

Other participatory practices located within the state sphere can incorporate or generate public interest and involvement in urban, project or site-level plans. Planning or design charrettes (public meetings undertaken for a concerted and foreshortened effort) with greater or lesser input from citizens have been popularized as a means of stimulating public interest in practicable, time-limited and cost-effective ways; the organizational rules being that they should be inclusive, start with a blank sheet and provide just enough information, and that the design or image acts as a contract (Roggema, 2013: 16). To the extent that these rules can act to limit citizen involvement, they might be thought of as experiments restricted to the intersection of state and club interests depicted in figure 2.1 (chapter 2). Here the 'best charrettes are intensive enough to build and hold the interest of participants, and just long enough to permit people to make considered judgements about the plan' (Hack, 2018: 125).

However, deliberative plan-making processes can become prolonged into major exercises spanning years. Deliberations over plans for the 26-acre Prudential Center site in Boston at the junction of four districts took 200 meetings and two years to reach agreement (Hack, 2018: 122). The time frame and participatory process here are not dissimilar to then Mayor (now President) Jokowi's planned relocations of street vendors in Surakarta, Indonesia, which took over 100 meetings to come to fruition (Bunnell et al., 2013).

Architectural or design competitions for particular development sites are used extensively by city authorities in Scandinavian countries for major public buildings or projects. They are not without their complexities, and the different actors involved may have contradictory interests (Tolson, 2011). One problem with such competitions is that 'if the program for the site is not precisely defined, it is difficult to conduct a definitive competition, since the jury will be forced to compare apples and oranges' (Hack, 2018: 128). More fundamentally, of course, these competitions may foreclose debate regarding alternative proposals. The thought that urban planning routinely makes use of counterfactuals (Hoch, 2019: 20) could be thought questionable when it comes to processes of statutory planning. Thus, where local government urban and land-use plans have gained public interest, it is usually in the form of outright opposition to rather than the formulation of alternatives to particular proposals. Davidoff's (1965: 333) suggestion that those critical of plans produced via statutory processes and associated consultation be charged and licensed with producing alternatives remains valid, as good examples of rigorous and viable alternative plans are hard to come across. One exception here is the detailed *Plan for Melbourne* (volumes 1–3) orchestrated by activists Ruth and Maurie Crow and published by the Communist Party of Australia as an alternative to plans being prepared at the time by the Melbourne and Metropolitan Board of Works. Inside a rather conventional cover for the day were proposals that favoured public transit rather than freeway building; higher-density living rather than suburban sprawl; more decentralized settlement planning rather than concentration on the state capital; and less negative, restrictive land-use planning and more design planning of whole integrated communities. Several of these ingredients have since become orthodoxy, but they were attended at the time by proposals for state control of 'all big manufacturing, extractive, commercial, recreational, administrative, or "development" concerns as well as adequate finance for them' (Communist Party of Australia, 1970: 21) that remain unorthodox.

Communicative action and urban planning

Public participation remains a focus for urban planning within the apparatus of representative democratic institutions. The good practical intentions of theories of communicative (Forester, 1993; Innes and Booher, 2010), collaborative (Healey, 1997) or deliberative (Forester, 1999) forms of public participation in statutory urban planning processes are clear but highly idealistic. These theories suggest that open dialogue among all interested parties can, if handled sensitively and held over as long a time as is needed, arrive at 'ideal speech' situations in which intersubjective understandings and open, undistorted communication occur, leading in turn to consensus on courses of action. It is hard to know whether any planning deliberations have approximated the theoretical ideals, and there is always the tautological explanation to hand that those many instances that fail to live up to ideals must not have been conducted according to the theoretical ideals of the communicative processes.

The difficulties of reaching consensus rub up against the practical limits of the time frames open to professional planners involved with typical forward planning or development application decisions. The value of the sorts of consensus that might be achieved from collaborative, communicative and deliberative processes of urban planning might be questioned where consensus barely exceeds the lowest common denominator of what is acceptable to all parties. In the case of the 'Gatwick Diamond' inter-municipality sub-regional planning for population and employment, undertaken in the 2000s in the UK, statements ultimately included no specific numbers of houses to be built or locations where they were to be built (Valler and Phelps, 2016).

More fundamentally, the common situation that confronts urban planners and the plan-making process is one of a 'dark side' of participation in which asymmetries of power are actively mobilized by actors (Flyvbjerg, 1998). This includes the deliberate distortion of communications, including the likes of dramaturgical behaviour or acting (Phelps and Tewdwr-Jones, 2000). Disagreement can run deep in global south or polarized or conflict-ridden city contexts subject to violent struggle, where collaborative or communicative statutory urban planning processes can bear insubstantial fruit or, worse, be seen as part of the problem (Bollens, 2002; Yiftachel, 1998). As Brand and Gaffikin (2007) detail, parties in Northern Ireland simply could not reconcile their views behind the rhetoric of collaboration. There, strategic spatial planning involved:

> public engagements that were both intensive and extensive, involving major civic conferences, focus groups, household questionnaires, youth forums and community workshops. But they did not commit any agency to anything definitive. While such warm words are preferable to chilled sectarian invective, they do not tie any side to any precise obligation for specific unpalatable change; and a currency that holds little cost also holds little value. (Brand and Gaffikin, 2007: 300)

Dissonance (rather than consonance or consensus) can also be quite deeply ingrained around long-standing urban planning dilemmas in some of the wealthiest and generally conflict-free parts of society, such as Oxfordshire in the UK (Phelps and Valler, 2018).

The new technologies of smart urbanism present some potential to democratize urban planning if they reach mass audiences. This is urban planning as 'crowdsourcing'. Indeed, taken together with the outsourcing of elements of expert data gathering and decision making to private sector consultants that is now commonplace in local government planning departments across the global north, urban planning may increasingly be in the service of the state but formally outside of it. There are parallels with the sorts of mass observation exercises undertaken in the UK (from the 1930s to the 1950s and again from the 1980s) – the point being that while these remain a fantastic informational resource, it is unclear how exactly they fed or are to feed into visionary urban planning and policy decisions.

Two contrasting cities stand out as pioneers in the use of new technologies to provide an urban-level platform for participation in government decision making. The Helsinki case can draw on a widespread and high level of trust in public officials including planners, to such a degree that mass responses to 'matter-of-fact' consultation on and participation in planning issues are routine. In the Barcelona case, the cultural tradition that participation can draw upon is completely different. Here, the ethos being created for public participation draws on a long history of grassroots demands for the right to the city from the time of the Franco dictatorship. Here participatory methods based on the city's digital platform are significantly oriented to the idea that the city needs to be affordable, not just smart.

Planning without statutory planning: agonism, co-production and radicalism

If communication, collaboration and deliberation are highly worthy aspirations that remain a valuable part of the urban planning imagination, contestation cannot be escaped. For Bhan, Srinivas and Watson (2018), fundamental contestations over the terms and directions of social change, theories of knowledge, modes of disciplinary and professional praxis and their institutional locations are very much a part of the urban planning imagination. In light of the limitations of 'orthodox' and communicative, collaborative and deliberative approaches to fostering public participation within statutory planning processes, then, more radical understandings of and methods for promoting participation in the planning of cities have emerged.

The emphasis in academia and practice on defining, training and protecting the status of professional planners seems out of place in a world where the majority of marginalized peoples seek ways of taking urban planning into their own hands, whether at the level of citizens, individual neighbourhoods or even across whole cities (Miraftab,

2009). The insurgency of citizens is registered in a fundamental sense of belonging to particular cities, not to nations and associated imagined communities, and is not bound by any of the shackles of colonial identities (Holston, 2009; Holston and Appadurai, 1999).

In theory, this co-productive and radical urban planning potential exists in the global north, since, as Alexander and Gleeson (2019: 96) predict, 'If conventional representative democracy in advanced capitalist societies is unable to accommodate the degrowth imperative by virtue of politicians and dominant institutions being locked into the growth model, then it follows that the emergence of degrowth in the suburbs will have to depend on a post-capitalist politics of participatory democracy.' Much of this radical urban planning potential in the global north remains as yet theoretical or restricted to limited experiments. It may be more widespread in the co-productive possibilities for 'citizen science' – those of mass collection of data – precisely in those instances where any data, let alone big or real-time data, is missing, as was the case in the wake of Hurricane Katrina in New Orleans (Thompson, 2017). Nevertheless, some of the advocacy for communicative, collaborative and deliberative planning has given way to theoretical approaches that emphasize the conflicts that are inherent and indeed valuable to urban planning deliberations across the global north. Preston (2009) reports on the findings of the 'Planning at the Edge of the Millennium' report produced by the California Planning Roundtable in 2000, which noted how urban planning had become *both* more collaborative and more adversarial. Thus, 'agonism could be said to be the ethos of a democracy respecting the legitimacy of difference and interests through public participation. Public planning should ideally be a place for strife about legitimate opinions and meanings on the road towards reasonable and commonly agreed solutions or consensus-building among mutual adversaries' (Pløger, 2004: 72). If anything, theories speaking to the agonism and disagreement inherent in urbanism as a way of life (Pløger, 2004; Sager, 2012) have remained yet more removed than theories of collaborative, communicative and deliberative urban planning from normative questions of what should be done. While individual planning processes or decisions almost routinely provide evidence of fundamental disagreements among interested actors (Mouat et al., 2013), it remains unclear what, if anything, can or should be done by the 'ordinary' local government planner or for the 'ordinary' citizen, both of whom are increasingly unrecognizable in global north theory and statutory planning processes.

There is a need to provide urban planning arenas for conflict to be expressed and not buried. This is especially the case in conflict-ridden contexts such as in Israel-Palestine and Northern Ireland, where statutory planning has often been deeply implicated in marked inequalities in claims to property and land, development opportunities and access to urban services (Yiftachel, 1998). Moreover, conflict and contestation may be no worse in their urban planning outcomes than collaborative or communicative approaches in these contexts, where, as we saw above, the latter appear to have generated largely empty words (Brand and Gaffikin, 2007). For Bollens (2002), polarized and conflict-ridden cities present an informative window through which to glimpse some of the same complexities now becoming more apparent in cities in North America and Western Europe. Research from the Decidim participatory planning laboratory in Barcelona in Spain suggests that conflictual statements at the outset of participatory urban planning exercises may promote better discussion than consensual statements (Aragón et al., 2017).

To date, the main examples of, and demand for, such opportunities have arisen across the global south. The groundswell of interest in radical alternatives to statutory approaches to participation stems from some of the contradictions of local government urban planning found in the likes of African cities: 'It is not just a case of bad and inappropriate planning, but also the absence of good planning that characterizes the African city. There is an absolute shortage of qualified built environment professionals and the brain drain further dilutes the small pool of experienced planners' (Parnell et al., 2009: 23). The picture of the statutory planner that emerges is diverse and spans the range from the promoter of the sorts of clientelistic relations we associate with corruption or the selling of planning permissions or rezonings, to a co-opter and de-radicalizer of citizens, to the planner powerless in the face of both land grabs by the wealthy and the direct actions of citizens resistant to evictions. In this regard, professional local government planners are themselves exposed to contradictory behaviours, since 'poor communities will often shift between clientelism and rights-based citizenship claims, using apparently contradictory discourses opportunistically (a strategy of "tactical bricolage") ... Such relationships go against the grain of both collaborative planning and co-production, yet they cannot be either wished away or policed' (Watson, 2014: 70). Here we have a sense of the city as something unplannable (Roy, 2009a) viewed through the lens of received global north textbook wisdom.

Why, asks Pieterse (2008: 111, original emphasis), 'can we not see slums as *constitutive* of the city as are, say, the suburbs, or the central business districts, or the luminous commercial retail spaces that serve to anchor middle-class existences in the postmodern city?' The inability of formal statutory urban planning processes to see major parts of the city leaves a vast reservoir of urban planning knowledge left untapped, since 'unless the complex, dynamic, high improvising and generative actions of the urban poor are acknowledged it is foolish to come to conclusions about what is going on in the city, or what may or may not work' (Pieterse, 2008: 3). If radical change is the existential core of city life, today it may coexist with a countervailing prudence and restraint found in planning by states and clubs. It is this that leads Pieterse to speak of the need for 'radical incrementalism'. Moreover, some of the global south citizen-centred urban planning radicalism, discussed further in chapter 6, is hardly irrelevant as a response to global north urban conditions.

In this connection there remains a need for an element of academic-led or -inspired incremental radicalism even in the largest informal settlements, such as in the *macrocampamento* Los Arenales in Antofagasta, Chile, where, despite the 1,000 households and multiple representative committees, citizens have lacked organizational capacity and have remained unempowered in their engagements with the state (Arias-Loyola and Vergara-Perucich, 2021; Vergara-Perucich and Arias-Loyola, 2019). Here the experiment at the interface between club (in the form of the local university) and citizen actor was intended to generate a sense of empowerment. Its own failures as an experiment reveal the limits of statutory and club urban planning imaginations. Nevertheless, if we learn from failure, then states and clubs in their engagements with citizens will need to leave enough resources, time and space for failures that may prove invaluable to the future of cities.

While there is a lot for urban planning in the global north to observe, absorb and learn from in experiences in cities across the global south, there are some important lessons to be *relearned* from the past of urban planning activism across the global north. William Bunge's (2011 [1971]) urban planning activism centred on the Detroit neighbourhood of Fitzgerald is one point of reference for a discussion of

Figure 5.2 Los Arenales, Antofagasta, Chile
Source: Martin Arias

the possibilities, though it is also a reminder of the limited freedoms for academic engagement with local communities. Somewhat later, in the UK, the bipartisan Community Development Projects (CDP) from the late 1960s to the late 1970s brought together citizens with university academics and statutory planner activists. The work of the CDP is now largely forgotten, but the twelve CDP groups, formed in multiply deprived communities, pioneered insightful diagnoses of urban problems and ideas that were ahead of their time and remain relevant today. Local government staff were to act as intermediaries, but outgrew their intended role as better coordinators of existing policy support to communities to become 'organic intellectuals', prompting criticisms of the programme's impact on community development practice (Banks and Carpenter, 2017). Much has changed in the UK since this time, and while it remains a challenge for universities and university academics to animate public participation with respect to urban planning, initiatives such as Newcastle City Futures demonstrate some of the possibilities (Tewdwr-Jones et al., 2019). The initiative of universities may be all the more valuable given the time it will take for senses of community to gain popular and inclusive support from particular geographical neighbourhoods within the UK's localism experiment, as the example from Stamford Hill in the London Borough of Hackney reveals (Colomb, 2017). In the case of Stamford Hill, two competing community groups have had their attempts to be designated the neighbourhood planning lead organization turned down twice, because both have failed to develop an urban planning imagination beyond their narrow club interests.

The possibilities for greater and more meaningful participation are signalled in ideas of 'co-production' in urban planning processes (Mitlin, 2008). Perhaps the most celebrated instance of citizen–state co-production in urban planning processes is the 'participatory budgeting' approach pioneered in Porto Alegre, Brazil, in a context of financial austerity, which I discuss in chapter 7. Elsewhere, other necessities have been the mothers of invention. In Japan, the impacts of a natural disaster drove *machizukuri* community-based planning approaches. Co-production has some of its own contradictions for some, since this form of participation can become a tyranny in the developing world.

Co-productive urban planning methods are subject to a feature of all urban planning acts, namely unintended consequences. Proponents of participation and co-production 'can no longer juxtapose the alleged benefits of bottom-up, people-centred, process-oriented and "alternative:" approaches with top-down, technocratic, blueprint planning of state-led modernization. The mainstreaming of participatory approaches to development ... has helped to blur these real divisions' (Hickey and Mohan, 2004: 4). Participatory approaches have a richer and longer history than is often appreciated but are essentially localist in orientation, often failing to acknowledge 'the material basis for identity, exploitation and politics' (Hickey and Mohan, 2004: 17). Moreover, strong tendencies to normalize human arrangements mean that new institutions and practices come to be invested with old understandings and attitudes (Cleaver, 2004).

Finally, in all of our deliberations about who should plan our cities, 'there is no sense in which formality precedes informality, any more than the state precedes the city' (Dovey and King, 2011: 12). It is hardly the case that the informal planning of citizens is any less fundamental to the formation of cities than the formal planning of states and clubs. Indeed, citizens of informal settlements have increasingly sought to engage with some of the most internationalized planning agendas – such as the UN-Habitat's Sustainable Development Goals (SDGs) – on their own terms, as in the case of Rio de Janeiro's favela communities.[3] The large, set-piece Kumbh Mela temporary city of worship that comes into being only once every twelve years at four different riverbank sites in India provides evidence of a form of urban planning that exists largely outside of the state system. For some at least, the experience of the Kumbh Mela shows how 'limited government, intentionally working to provide a platform for other actors to provide services, can accomplish a substantially better outcome than a resource-constrained government trying to do everything itself' (Khanna and Macomber, 2015: 343). Here the inclusiveness of temporary urbanism surpasses some of the statutory planning operative on permanent cities. There may be some lessons to be learned from this temporary city built in the form of the most inclusive of club projects.

Mixes of methods for new urban planning imaginaries

The severe limitations of expert-led, top-down statutory urban planning practised in the name of representative democracy, coupled with the vigorously emerging demand from clubs and citizens to actively co-produce in urban planning processes, appears to presage a future urban planning without planners as we have come to know them; a future of urban planning without the professional (local) state planner. Much of the stock of urban planning theory and practice literature and associated thinking has come to speak to the generalist-expert, professionally accredited local government planner working in the service of the public interest. Yet this version of the planner increasingly cuts a lonely, undermined or bureaucratically remote figure. In many if not all contexts, most urban planning takes place outside local government bureaucracies in the everyday actions of citizens, in consultancies, in NGOs and in clubs of limited interests.

Given this, and partly thanks to the efforts that fed into the UN's formulation of its SDGs, the city can be and has been recast as a laboratory for experimentation (Barnett and Parnell, 2018; Bulkely and Castan Broto, 2013), but one in which traditional planners are not the most agile or innovative when compared to either citizens or club actors. State planners need more than ever to have the skills and aptitude of their fellow built-environment professionals – the architects – to seduce, to 'resonate with the values, perceptions and particular needs of key actors' (Healey, 2007: 192). Indeed, revaluing planning entails refinding some of the imagination of urban planning – in the form of artistic influence, the diffusion of practical innovations and the art of persuasion – as it was practised and spread in the late 1800s to early 1900s (Sutcliffe, 1981).

Planning's artistry

Urban planning has been visionary. Past plans adorn the walls of many a university planning school and are little short of works of art. Their beauty has sometimes beguiled for the wrong reasons: those of dispossession and money making. Other plans have seduced for all the right reasons, painting an image of the planner as someone sketching the narrative and visual arc of a better life in a better place. It remains the case that a good visual is perhaps the planner's greatest weapon (Garvin, 2009) – this much is clear from the way planning ideas continue to travel far and wide, as I discuss in chapter 7. Daniel Burnham, the author of the 1909 *Plan of Chicago*, famously proclaimed: 'Make no little plans; they have no magic to stir men's blood and will not be realized. Make big plans; aim high in hope and work, remembering that a noble, logical diagram once recorded will never die, but long after we are gone will be a living thing, asserting itself with ever growing insistency' (quoted in Garvin, 2009: 34). Of course, visual images including plans fail to capture the popular or political imagination. Geddes' rhetoric was not matched, it seems, by his visual imagery, it being 'immediately obvious how impenetrable and unhelpful his "thinking machines" were either as sources of persuasion or as efficacious educational tools' (Meller, 2005: 49). No matter how good or how visually arresting a plan is, it must therefore appeal to citizens and politicians if it is to have traction. One example of the dire meaning of a plan for local communities is Colin Buchanan's linear grid city for south Hampshire in 1966, which failed as a map for the strategic growth of the area despite appearing at the height of political and popular interest in the white heat of modernity in the UK (Phelps, 2012a).

'Cultural representations of cities can enrich our understanding of urban environmental change' (Gandy, 2014: 18). And to the extent that planners – citizens, those in the service of clubs or the state's professional planners – can represent the culture of cities visually, then they can bring others with them. Something of the vast array of visual tools and methods associated with their use and interpretation is coming to light (Rose, 2001). Thus, while the traditional land-use plan may be less fit for purpose in a fast-changing world, visual methods as a whole appear likely, if anything, to retain or even increase their importance to planning practice, though care is needed in asserting any unilinear historical trajectory here (Rose, 2001: 9). It is clear that photographs and films of plans, planners and the planning process can reveal much of the motivations of the various actors involved. They may be powerful and disturbing reminders of what has been done or intended in the name of planning. Equally, they can operate as a way of broaching the need for plans, drawing a wide variety of actors into planning processes, and doing planning (Tewdwr-Jones, 2011). Visuals can just as often remind us of paths not taken in urban planning decisions and of the alternatives that seem so unimaginable from the present.

Experimentation

Much recent academic and, to an extent, practice reportage alights on the idea of the city as a laboratory for various policy experiments. This has usually been cast in positive ways to emphasize the instigation and orchestration of socially useful experiments and alternatives. Yet for the most part the image of the professional local government planner is of somebody removed from such experimentation. However, the professional local government planner might be recast as one of a large number of actors involved in the creative process of city making. This is an image of the professional planner unburdened – as far as this is made possible within the state system – of some of the need to neutrally facilitate, administer and deliberate among competing ends. It is an image of the professional local government planner licensed to draw on his or her training, practical experience and imagination as an advocate, consultant or mentor for the making of good or better places within planning processes. This is now vital in an age when the due consideration of alternatives, or the counterfactual of what might be in the absence of a particular plan as a key element in any planning decision, has been so sorely lacking. In an age of TINA ('there is no alternative'), the professional local government planner might yet emerge as one emboldened to encourage and facilitate as many alternative plans as possible – city-, neighbourhood- or project-level – by whichever actor, in whatever stage of completion, to whatever degree of rigour is possible, as a vital input into due deliberation of what might constitute better urban futures, in contrast to the production of single alternatively glossy or indifferent places.

It is rare to think of planning in this way, but plans, visions and policies can have political currency that planners themselves can be aware of. Even if not readily realized, these are a store of value to the extent that they can induce others to act in the building of the urban environments in which we live. Here the professional local government planner takes his or her place as an equal alongside other built-environment professionals – architects, valuers, developers, landscapers, realtors – as a mediator in the building of the urban environment.

Planning rhetoric

Patrick Geddes' influence within and beyond the nascent planning profession doubtless derived from his rhetorical flourishes or the sort of 'culture jargon' commonly deployed by architects. Among the built-environment professions, planners are by now among the least celebrated, and some of this lack of recognition must surely derive from the noticeably inhibited use of language made by professional local government planners in their plans and presumably in meetings with elected representatives and the public.

One of the key skills of the professions or would-be professions that have been carved from, and continue to operate within and *distinguish* themselves as intermediaries within, markets is the ability to use language in ways that regenerate cultural capital. Compare, for example, the cachet attached to the newly created professions of management consultants (McKenna, 2010) and international investment arbitrators (Dezalay and Garth, 1996) with the contempt in which professional urban planners in the state sector of a liberal market economy are sometimes held by politicians, citizens and club actors alike. If cities are the physical embodiments and accumulated stores of culture, then the professional planner can yet be our informed guide to and guardian of them. The planner, if he or she can find the rhetoric, 'evidence, argument, and persuasion' (Majone, 1989), is at least the equal of the other built-environment professions as the natural guardian of the public interest that inheres in the city. The notion of the public interest has become increasingly difficult to define, yet it may be something measured in the accumulated experience of the professional planner (regardless of which citizen, state or club interests are immediately being served), an experience that is rarely allowed to be unleashed in creative ways. Other industries are replete with expertise that is valued as being part art, part science, gleaned as much from practice as from textbooks and exercised in the policy process (Majone, 1989).

Conclusions

The urban planning imagination involves more than the collation and processing of information that Mills (1959: 114) lamented as accompanying the bureaucratization of society, including academia. The range of tools or methods of urban planning reviewed here give some sense of how the urban planning imagination includes but exceeds quantitative data and trends, and invites reflection on the past and the active shaping of futures. 'Thinking in numerical calculations, stories, and images simultaneously and iteratively is essential to inventiveness, effectiveness, and persuasive collaboration' (Hopkins and Zapata, 2007: 1) and provides an indication of the breadth of urban planning's methodological imagination.

Yet for much of plan-making activity 'there has been a heavy reliance on the so-called "creative leap", a mysterious jump from survey findings to plan proposals' (Batey, 2018: 57). There is a need for the creative leap made by expert planners to be replaced by or supplemented with something else, less narrowly expert and more inclusive. It is nevertheless worth considering whether this creative leap is a problem that has hampered urban planning or an opportunity no longer seized by or even open within the planning profession. Other professions – architects and management consultants – have been able to appeal to such creativity as part of their professional aura.

The urban planning methods I have considered here have some common problems. All can, for example, be associated with a measure of decision-making sclerosis or paralysis. I have argued here that the repertoire of methods – solitary and technical, but also participatory and visionary – is a source of creativity that ought to be recognized and valued by the public and politicians. The appeal to an urban planning imagination might seem oxymoronic, as by now statutory planning and planners have perhaps become synonymous with a bureaucracy of regulations and administrative procedures. A further problem is that professional accreditation runs up against educational freedom. One recent comparative study of the different professional planning accreditation standards in Australia, the UK and the US revealed that 'none of the documents explicitly set out the role of the planner as having the ability, intellectually, to develop ideas and to set new agendas as a professional. Thus, their potential as change agents is not explicitly facilitated or sought' (March et al., 2013: 240). It is time to recapture some of the imagination of urban planning, its verve and experimentation across state, club and citizen actors.

6 Comparisons: what are the global variations in planning?

Introduction

While the majority of us now live in urban societies that are ever more connected globally, urban planning remains highly differentiated in both the disciplinary training and thought that goes into it and the day-to-day practice that comprises it. In this chapter I consider some of the contrasts among systems and cultures of urban planning. The comparative urban planning literature is modest for the reasons encountered with all comparative research: the linguistic skills required for comparison and the time and sensitivity needed to appreciate the cultural and other contextual shapers of urban planning (Nadin and Stead, 2013). The state is the key reference point in much of the extant literature, which, as a result, remains prone to 'methodological nationalism'.

The view that urban planning cannot 'be understood as anything other than an end-product of political, and administrative and legal forces that shaped a country's constitution' (Booth, 2011: 16) has been prevalent in the literature. However, Rose (1984: 35) broadens the search for the forces shaping urban planning systems to the 'history, geo-politics and cultural norms of that country'. As such, the influences upon urban planning systems and cultures are usefully considered from disciplinary perspectives outside of urban planning. I elaborate on the distinctiveness of subnational planning as a part reflection of national economic development or welfare models. Discussion here reveals some of the connections between, and processes promoting convergence among, national planning systems and cultures, covered in the next chapter.

Different traditions are a part of the vitality of the urban planning imagination. Indeed, global variations generate the tendency for actors to want to exchange ideas and practice. Contrasts and comparisons feed the planning imagination, prompt the generation of theory and methods, and add to the stock of wisdom that urban planning practice can leverage. Comparative analysis may often be motivated by a need or desire to improve practice (Booth, 2011: 15), and among the practical benefits of comparative analysis are understandings of the effects of policy interventions and evaluation of policy processes.[1] It is the vast and varied global 'market' for urban planning wisdom, ideas, concepts and methods which suggests the intrinsic and practical value of urban planning. It is a market I cannot do full justice to here, especially in connection with the continuing need to move discussion away from European and North American reference points.

Comparative approaches

'Comparisons across systems that differ in history, culture, legal tradition, and other important contextual aspects are inherently difficult' (Wolman, 2008: 87). There are a number of different disciplinary traditions that offer methodological insight into comparisons of planning systems and cultures. They reveal the continued diversity that exists in a world drawn closer together as a result of greater physical and virtual mobility. Comparative methods across disciplines appear to have converged to recognize – as with practice-centred international agendas such as the UN's SDGs – the city as a more appropriate unit of analysis than the nations in which they remain nested.

Politics, political and administrative science

Politics, political science and policy studies are noted for their highly parsimonious comparisons with respect to politics and formal governmental forms and outcomes nested within nation states. While urban studies taken as a whole emphasize the importance of adopting 'a holistic and context-embracing approach to understand the contemporary city or the historic trajectory of urban development' (Pierre, 2005: 447), the study of urban politics has tended to adopt more parsimonious designs that compare and contrast cities along a limited number of dimensions and consider their autonomous effects to be operative within their respective nation states. Something of the potential complexity of national systems of local government is indicated in Wolman's nine distinct dimensions along which comparisons could be instructive – most of which have connections to urban planning.[2]

Nevertheless, even the study of urban politics 'seems to have embraced complexity and richness in context at the expense of parsimony' (Pierre, 2005: 449). This is partly because urban rather than national politics may be more amenable to comparison, as there is simply a larger universe of cases to explore and with which to generate better comparisons. It is in part a response to the recognition that some of that wider local context is gained by including a range of actors and processes that exceed city government under the label 'urban governance'. The distinctions between, and explanatory status of, government and governance remain a source of debate.

As 'the significance of governance and politics within cities and regions grows, comparative urban analysis has an increasingly prominent role to play in the ... study of politics and policy' (Sellers, 2005: 419). Looking to Europe, some of the complexities of comparing urban planning reflect the multi-level nature of governance in the EU (Marks et al., 1996), in which national- and city-level processes are not always distinguished adequately (Sellers, 2005) and in which the autonomy of distinctive city-level processes (including those bearing on urban planning) can easily be overstated (Harding, 1997). I cannot do full justice to these multiple levels of influence in this chapter alone, but something of their influence emerges across the discussion in this and the next chapter.

Geography and history

When it comes to comparative analysis, the two disciplines I emphasized from the outset as natural companions to planning part company. As we saw in chapter 2, historical analysis focuses on the long-term macro-processes and short-term micro-processes described by Tilly (1984). Geographers have tended to move away from some of the parsimonious comparisons typically designed by economists and political scientists, on the one hand, and the search for a few common structures underlying local variations, emphasized by historians and political economists, on the other hand. Geographers have thus moved away from comparative methods designed to generate universal explanations of phenomena towards those that are more revealing of the peculiarities of places – in the form of individual case studies of particular national planning systems or city-governance arrangements (Burawoy, 1998; Flyvbjerg, 2006) – and more encompassing variations on common themes.

'Classification is often neglected or, worse, derided as a simple descriptive exercise' (Wolman, 2008: 89), and something of this sentiment appears to be strong in geography, where classifications have come to be regarded as reductionist. Part of the problem here is that they are rarely rigorously explained or deployed – this includes the classification produced in table 6.1, which I have compiled as one way in which some of the comparative complexities in which urban planning systems and cultures are nested might be reduced.[3] Looser comparative research designs have emerged in urban studies, partly with the intent of drawing attention to otherwise neglected, 'ordinary' cities or alternative reference points for the development of theory (Robinson, 2006), and partly to reflect geographical sensibilities that speak to the many relations that transcend the divides typically erected in classification schemes (Robinson, 2011). Part of the value of such looser methodological designs is that juxtaposition of cases is used as subtraction, not confirmative addition, of a succession of cases, to better understand urban processes as complex and varied in the way they work out locally and in order to glimpse possible urban futures (Jacobs, 2012: 906). The same relevance of subtraction as a methodological strategy for analysis is apparent in the juxtaposing of cases within individual cities (MacFarlane et al., 2017).

In these developments across the disciplines we can observe a movement beyond 'methodological nationalism', seeking out the contributions of citizens and clubs to the evolution of nation-state-centred contrasts in a world in which we can glimpse partial commonalities. For now, it is enough to note that much of the variation found in urban planning systems and cultures remains a product of nation states and statutory planning.

National planning systems and cultures

Much of the literature comparing urban planning systems and cultures specifically focuses on the national scale and derives from the birthplace of the nation state: Europe. Europe is notable not only for the number and variety of nation states in a continental space that elsewhere might be home to just one or two, but also for the projection of national influence (including in the sphere of urban planning) internationally. The main classifications of national urban planning systems have their limitations, but they reveal just what a varied urban planning world we continue to live in. Some indication of this is provided by the facts that there were 229 different types of plans in operation in the thirty-two OECD countries, and in excess of 100,000 land-use plans in existence across these countries (OECD, 2017a). The variety that exists among these wealthiest of nations should alert us to the wider variety that exists in urban planning systems.

Notwithstanding the general definitions of urban planning offered in chapter 1, it is clear that such definitions 'represent a meaning specific to the ... state (or perhaps even region) where they are used, and are not directly transferable from one situation to another. This is true even where the same term is used in the same language in different countries' (CEC, 1997: 23). The European Commission's use of the umbrella term 'spatial planning' 'encompasses elements of national and transnational planning, regional policy, regional planning and detailed land use planning' (CEC, 1997: 24). The comparative literature that exists speaks loudly to the thought that urban planning takes its cues from its legal and administrative context (Newman and Thornley, 1996).

Planning systems: formal contrasts in law and administration

Although Davies et al. (1989: 439) drew a contrast between the flexibility said in principle to characterize the UK and the rigidity that in principle inheres in the different legal traditions across mainland Europe, they also noted that contrasts were, in practice, never as clear cut (Davies et al., 1989: 439). At the time, with the 1992 completion of a European common internal market, comparisons between the UK and continental urban planning systems prompted thoughts of convergence among national planning systems. Berry and McGreal (1995) identified a common trend for European cities to become more entrepreneurial and planning systems more pro-development oriented. Healey and Williams (1993: 704) were prompted to muse how 'one of the most interesting questions with respect to convergence tendencies in European planning systems is whether legal systems derived from Roman Law (and in Southern Europe the Code Napoleon), will continue to dominate or whether the British

Table 6.1 Informal contrasts between types of planning systems

Model of welfare or national economic development	Exemplar countries	Planning imagination	Leading actors	Methods	Geographical sensibility	Temporality
Liberal market	Australia, UK and US	Administrative efficiency	States, clubs and citizens	Increasingly ad hoc and project-based with residual comprehensive efforts	Scalar giving way to networks and flows: competitive capture of economic flows and positioning of cities within networks	Churn in the short term, lack of long-term planning
European social welfare	Majority of mainland Western European nations	Comprehensive and integrated	States and citizens	Comprehensive and integrated with hierarchy of plans and elements of redistribution	Scalar: social cohesion of city, region and national and supranational territories	Gradual erosion of longer-term planning
Developmental	Japan, South Korea, Singapore, Taiwan, China	Economic development and industry-sector-led	States and clubs	Strong long-term industrial and infrastructure planning, weak reactive urban planning	Scalar: export competitiveness of nations	Long-term objectives increasingly exposed to fiscal crises of the state
Predatory	Some Latin American and sub-Saharan African nations, Indonesia.	Non-plan opportunism	Clubs	Plans as incremental after-the-event legitimation	None explicit	Short-term private gains trump medium- or long-term urban planning

Model of welfare or national economic development	Exemplar countries	Planning imagination	Leading actors	Methods	Geographical sensibility	Temporality
Post-socialist/ transition economies	Russia, China, East and East-Central European nations	Mix of industry-sector-led planning legacy with variety of liberal market and European social welfare urban planning elements	States	Urban planning hampered by legacy of sectoral planning	Scalar: gradual emergence of stronger sense of territorial planning at city, regional and national scales	Variety from major discontinuous systemic change to gradualism of planning experiments
South Asian democracies	India, Pakistan, Bangladesh	Formal urban planning as 'victim' of elite and grassroots informality	Clubs and citizens	Plans as incremental after-the-event legitimation or new master-planned development	Scalar: contested spoils of 'slum' redevelopments, major urban extensions and new towns	Short-term club plans drawn out by longer-term contestation by citizens
'Born globals'	Middle East and Eurasian states, China (Dubai, Nur-Sultan (formerly Astana), Shenzhen)	Pick and mix of influences and practices	States and clubs	Visual imagery and symbolism as urban planning	Networks, flows and virtual: capture of niche symbolic economic flows and positioning of cities within real and virtual economy networks	Medium-term master planning of consumption-oriented and investment-seeking cities

administrative model will prevail. Both ... have proved adaptable' (Healey and Williams, 1993: 704). Such adaptability may ensure that post-Brexit divergence between urban planning in the UK and the EU is less than might be expected.[4]

In theory, the legal certainty offered in mainland European systems ought to be more suitable to the development of large infrastructure projects, major urban extensions or entirely new settlements, while the UK system was conceivably better suited to catering for incremental change within existing urban areas. Here again contrasts can prove misleading, since the likes of the Turnpike Commissions and in modern times the New Town and Urban Development Corporations demonstrate a long tradition in the UK of the creation of ad hoc bodies related to urban planning. What is said to be lost in the UK system is the predictability of outcome, either for routine small-scale development applications or large projects or for urban planning principles for which there has been a strong measure of consensus. This particular contrast – between the flexibility of discretion and the rigidity of legal certainty – may be overdrawn but is one that continues to resonate across a wider group of OECD countries today (OECD, 2017b).

Newman and Thornley (1996: 27) focus on both legal and administrative traditions found across Europe, noting how 'Planning gains its power through its embodiment in the legislation and regulations which form part of the legal apparatus of a particular country. ... The implementation of planning occurs through the administrative system which again varies considerably across the countries of Europe.' They identify five planning families in Europe. Law is deeply entangled with the demarcation of space and property (Blomley, 2014) and the nature of urban planning systems and their determinations (Blomley, 2017). As such, comparative law can be instructive at a global scale in understanding the character and scope of urban planning (Depaulle, 1922; Reitz, 1998). However, it is worth noting that this approach emphasizes differences in the formal legal and administrative operation of planning systems, including key tools such as land-use and spatial plans, as compared to the actual practice of planning (Nadin and Stead, 2008: 38): this despite the fact that 'the persistence of formal planning systems contrasts with the fluidity with which planning practices can change' (OECD, 2017a: 27). Moreover, despite good reasons for expecting major differences between unitary and federal state administrative systems, for example, at 'the subnational level, the practical consequences resulting from the differences between unitary and federal countries are less important than at the national level' (OECD, 2017a: 26). The differences that do appear to exist between the two administrative systems, at least at the local level, relate to 'soft' factors such as planning traditions or cultures and not to the 'hard' or formal legal and administrative factors (OECD, 2017a: 26).

Here again, the UK system stands alone, partly as a result of the specific way in which urban planning is rendered as land-use planning, partly because of the lack of any written constitution under which the jurisdiction and capacities of local governments are guaranteed, partly as a result of the recourse to case law and its pragmatism based on empirical evidence rather than abstract principles, and partly as a result of the element of discretion involved in urban planning decisions at the local level. The discretion found in the UK system is three-fold, since local planning officers have discretion to interpret considerations material to any planning application (including the local plan and central government policy), but elected representatives are not obliged to follow planners' recommendations, while the secretary of state can 'call in' individual local decisions.[5]

The largest family of urban planning systems in the EU at the time of Newman and Thornley's (1996) writing was the Napoleonic family, so called after the Napoleonic legal code. Following the French Revolution and Enlightenment thinking, countries here have written constitutions which attempt to clearly and completely spell out the rights and obligations of citizens in relation to government, and of the different tiers of government. In these countries, 'The aim has been to think about matters in advance and prepare a complete system of rules based upon the codification of abstract principles' (Newman and Thornley, 1996: 31). These nations feature a code of planning regulations and hierarchy of plans (zoning) adhering to these principles. As with the Germanic family discussed below, the high level of codification of urban planning presents a certain similarity with the urban planning systems in operation across much of the US, despite the latter's liberal market economy similarities with the UK. However, it is also worthwhile to consider the differences that exist within this family. These can in some instances be summarized in terms of the gaps between the rhetoric and the reality of urban planning. The 'statism' in law and practice found in France might be contrasted with the frequent inability of the state in Spain, Italy and Greece to exert authority, control and enforcement in urban planning matters despite the existence of the relevant laws. Where in France urban planning or the tradition of *aménagement du territoire* leads urban development processes, in Spain, Italy and Greece it is as likely to follow the reality of development patterns.

The CEC's distinguishing of the urbanism tradition characteristic of these Mediterranean states reflects some of this gap between rhetoric and reality in their codified systems of planning, as well as some of the different disciplinary training whereby the ancient historic cores of Mediterranean cities necessitate strong architectural expertise in the sphere of urban planning. It also reflects the way urban planning is swept up in understandings of the process of modernization and of the city itself (Leontidou, 1990). Thus 'the frontier of urban expansion is determined by popular land colonization and illegal building', and authorities have tolerated and even promoted illegal building for its contributions to easing social conflicts and as one way to garner political patronage (Leontidou, 1990: 19, 148). The urbanism tradition actually renders urban planning in the Mediterranean nations closer in some respects to that in Latin America than to other European counterparts. Such connections exist formally in the Laws of the Indies that operated across much of Latin America, in the informality which remains significant and visible in suburbanization as an approach to the city (rather than as an escape from it as commonly understood in Anglophone nations), and in the exceptional episodes, or crises, that shape planning policies and practices (Lingua and Servillo, 2014).

The Germanic family might be considered a distinctive branch of the Napoleonic family (Newman and Thornley, 1996) in that it too retains a highly codified set of urban planning regulations and a strict hierarchy of plans. One crucial difference is the status of regions or *Länder* within this system, which reflects the historic pattern of urbanization and continues to ensure both regional variations in planning in a context of strong national principles, and relatively balanced development across the national territory. Despite strongly enshrined principles of regional balance, urban planning in some cities, such as Leipzig, has had to deal simultaneously with the shrinkage of the central city and peripheral overdevelopment (Nuissl and Rink, 2005). The differences here are more subtle in that codification, while abstract, is driven by intellectual (rather than ideological) priorities. The German system has a place in the early transmission of urban planning ideas. German towns and cities were among the earliest to deal systematically with the need to plan for or zone urban extensions. As a consequence, urban plans and associated mechanisms (such as land readjustment) developed in Germany were an early source of inspiration in the US (Hirt, 2007), Europe and Japan (Faludi, 2015; Sorensen, 2000). The German case is also a prime example of the potential for comparisons of different national urban planning systems to obscure what are often complex amalgams of competing traditions existing *within* a single country. The Germanic family has contained a tension between the two traditions of *Städtebau* (town planning) and *Stadtlandschaft* (urban landscape), whose influence both domestically and abroad has waxed and waned (Kress, 2018).

Newman and Thornley highlight the separateness of a Scandinavian urban planning family, which has strong commonalities among the nations involved due to patterns of Swedish and Danish conquest. While these systems were strongly influenced by the German and Napoleonic traditions before the 1800s, they then began to gain some of the distinctiveness for which they are known today. That distinctiveness might be characterized as based on rational, pragmatic, socially progressive agendas based on strong environmental protections, which at one point in the history of the EU set the bar for an attempted upward harmonization of standards.

Finally, Newman and Thornley highlight an East European urban planning family. This remains a large family to speak of in the singular for a number of reasons. First, pre- and post-communist influences play unevenly into each national urban planning system. A major communist-era legacy here may be that of sectoral planning, which continues to affect the coherence of local governments' place-making capacities to a greater or lesser extent. Second, there is a slow and highly politicized process of adjustment of land and property markets. Third, the transition to a market economy coupled with decentralization of competencies to local governments appears to have fuelled aggressive inter-city competition for residents and businesses, resulting in an oversupply of land for development (Wagner, 2018) which in some instances has been the subject of political decision making rather than urban planning per se. Fourth, there are various influences emanating from other European national planning systems in the form of technical assistance and the exchange of planning ideas, though these, as Maier (2012: 147) argues, may only be rhetorical.

Arguably, early typologies of European social models, governance culture and planning systems simply ignored East–Central/Eastern Europe (Maier, 2012: 138) and probably understated the internal heterogeneity of this family. So, for example, the national state is probably a more important reference point for citizens in these countries than elsewhere in Europe. And while municipalities enjoy wide competencies on local matters, including planning, they lack resources. It is with respect to this family of planning systems that we open up questions of where Europe and European influence begin and end.

Since these seminal discussions of the spatial planning systems of Europe, the membership of the EU has more than doubled to defy easy or consistent summary of either the commonalities or the differences in urban planning systems found there. The one common position across the member countries is that 'both the state and private individuals can own land. However, with few exceptions land ownership does not automatically confer rights to develop the land. In all cases the right to develop land effectively belongs to the state' (ESPON, 2018: 13).

Culture: the local personalities of planning

'Methodological nationalism' remains relevant for comparative analysis in a world in which new nation states continue to be formed and new capital cities built; older nation states are subject to powerful processes of decentralization and devolution; and, with the exception of a few countries (Germany, Spain), the status of subnational urban planning (whether regional, provincial or metropolitan) has never been strong. Yet the emergence of local planning systems and cultures is implied in the formation of nation states, and so the variety of urban planning systems and cultures continues to expand, not diminish.

The formation of nation states, whereby regions were politically folded or else militarily coerced into national territories, is precisely the same process that has been unwinding itself in the devolution and decentralization of nation states. 'New' nations have been carved off from or reformed from territories previously incorporated as nation states that have enjoyed periods of relative stability. For example, a prolonged paramilitary campaign largely ceased when a strong measure of governmental autonomy was ceded to the Spanish Basque country, which in turn has been credited with a 'Bilbao model' of urban regeneration. Protests continue for the full secession of Catalunya, to such an extent that in Spain several regional urban planning models might be thought to exist.

In the UK, the task of managing the national territory has seen institutions created and competencies and powers ceded to Wales and Scotland in connection with claims to separate nationhood. Decentralized administration had existed in Wales and Scotland from the early 1900s and, in the midst of violent protests in Wales in the 1970s, new regional economic development agencies were established. The gradual disuniting continued with the creation of a Welsh Assembly and a Scottish Parliament. Here, national spatial plans do indeed express aspects of nationhood, including partial realignments of territories and their international connections (Colomb and Tomaney, 2016).

The conflicts may be both more profound and more localized in nations colonized by major settlement or where imposed urban planning systems are now subject to a significant measure of contestation (the US, Canada and Australia). The complexities associated with recognizing and offering restorative justice to hundreds of first nations that existed in Australia prior to the arrival of and settlement by the British far exceed the challenges of territorial management of cities and regions folded into nations across Europe.[6] Here the process of dispossession was complete, with unique, customary and collective rights to land completely overlaid with a new system drawn up on the basis of the legal fiction of the taking possession of an unsettled land or *terra nullius*. The problem is that these native titles are being settled not only on just a fraction of the lands that would once have been native, but also by recourse to western legal concepts which themselves do not fit with former customary understandings of the rights to and use of land.

The ongoing difficulties that formal planning systems have in dealing with the functional realities of urbanization (discussed in chapters 2 and 3), as a result of their being trapped within largely unchanging local government jurisdictions, has prompted a measure of localization of planning in the form of fuzzy or soft planning spaces (Allmendinger and Haughton, 2009), now found across the UK. The same issues present themselves elsewhere in Europe, with

France's numerous small communes increasingly being tied together in *contrats de ville* designed to promote economic development and urban regeneration. The approach here – in a context of stasis in Europe – is different from that found in, for example, fast-growing China, where city boundaries have been redrawn with annexations of surrounding county jurisdictions to form truly massive new city-regions, with enormous latitude to accommodate future population growth and urban development (Zhang and Wu, 2006).

More fundamentally, and to pre-empt some of the concerns of the next chapter, the under-explored significance of urban planning cultures is signalled in the thought that culture itself remains *the* force driving exchanges of all sorts at a global scale (Bayly, 2000; Osterhammel, 2015). That is, 'culture lies at the heart of world development. Technical progress, bureaucratization, capitalism organization, states, and markets are embedded in cultural models often not explicitly recognized as such, that specify the "nature of things" and "the purposes of action"' (Boli and Thomas, 1999b: 17). Neither the culture nor the city-ness inhering in urban planning are new; indeed, they may be set to be reinvigorated in urban planning in ways that we can only dimly perceive at present.

Discussion of the four families of national planning in Europe identified in *The EU Compendium of Spatial Planning Systems and Policies* (CEC, 1997) are inflected with aspects of culture. Here something of the physical and civic inheritance of national planning systems comes to the fore. There are those national urban planning systems in which there has been a strong tradition of regional economic planning (represented by France and Germany). There are those that have been regarded as comprehensive and integrated (exemplified by the Netherlands). There is the UK system, which stands alone as one centred on land use. Finally, the southern Mediterranean nations (Italy, Spain, Greece) are those with a tradition of urbanism drawn from the need to protect the ancient and medieval architectural heritage of city cores. The cultures of the comprehensive-integrative and regional economic planning models have exerted a strong influence (in comparison to the land-use and urbanism models) on the Europeanization of urban planning, as discussed in the next chapter.

More recently, though, an interest in distinctly different cultures of planning has cut loose from considerations of the national legal and administrative bases of national planning systems (Reimer et al.,

2014). These bases ensure a measure of inertia, though laws and administrative structures and processes are revised quite frequently in the present era. Planning cultures may today provide the better part of the continuity apparent in urban planning practice. At the least, it seems that planning cultures articulate the legal and administrative bases of a given planning system with cues deriving from the wider societal context (Maier, 2012).

National planning culture has been defined as 'the collective ethos and dominant attitudes of planners regarding the appropriate role of the state, market forces, and civil society in influencing social outcomes' (Sanyal, 2005a: xxi). More concretely, it can be seen in the distinct norms, values and principles that underlie planning as it is practised (Dühr et al., 2010). Othengrafen and Reimer (2013: 1272–3) go further, to specify planning culture as a 'collective intelligible social practice' involving 'a number of incorporated and (implicit) routinised "recurrent regularities" about how to behave and act in specific situations ... It consists of beliefs, attitudes, ideas, norms, values, and behaviours that are "obviously valid" for members of the culture and guides the actions of members belonging to a specific culture.' Here the societal-wide value systems in which planning is embedded (for example, prevalent attitudes towards urban and rural landscapes, the primacy of private property and the legitimacy of state intervention) are important to consider. These impart something of the personality of planning. Some of these personalities reflect the basis of cities in nature, as in the origins of Dutch planning in the polder system and collective efforts to reclaim and defend land from the sea, or in the emergent specialization of Japanese urban planning in issues of disaster management. Arguably, the need to ensure economic survival first and foremost has coloured urban planning in Singapore ever since independence in 1965. In other instances, there has been no single driver – natural or otherwise – that imparts a clear personality to a planning system. The UK's temperate climate, stable geology and societal composition have resulted in a middling muddle or set of compromises when it comes to planning (Phelps, 2012a).

It is clear, however, that the common legal and administrative bases of national planning systems contain and become adapted to variable development pressures to produce different styles of planning *within* nations. In the UK, for example, Brindley, Rydin and Stoker (1996) identified six different planning styles across localities, namely 'regulative', 'trend', 'popular', 'leverage', 'public investment' and 'private management', reflecting particular mixes of policy goals, working methods and planning identities. Such local planning 'styles' have become firmly established in the UK since the formation of the modern town and country planning system in 1947. By virtue of sheer scale and diversity, federal constitution and vintage of settlement, some nations such as the US and China can hardly be considered to have single planning systems or cultures. There is a need, then, to move beyond some of the simplifications deriving from comparisons based on national legal and administrative arrangements to examine distinct local planning rationalities or cultures (Othengrafen and Reimer, 2013; Reimer et al., 2014; Valler and Phelps, 2018).

At the finer-grained local level, the different temporalities of planning (chapter 2) carry with them distinct cultures of practice. Planning cultures may inhere in, for example, particular frames that endure for decades across the plan-making cycle (Schön and Rein, 1994) and inform the initiation and maintenance of specific policies, the production of plans, and even accepted understandings of the prospects for individual sites. Further evidence of this is provided by Valler and Phelps (2018), who trace the effects of long-standing cognitive frames on sub-regional planning for growth in the southeast of England. To set against this, it is apparent that crises have intruded into national and local planning cultures and may be set to become more frequent moments in which cultures are disrupted (Reimer et al., 2014).

Models of welfare and national economic development

National planning systems and cultures can be placed in broader continental–regional contexts. Roy (2009b) summarizes some of the broad contrasts that can be found in the concerns common to groupings of countries in South and East Asia, Latin America and Africa. Here I absorb some of these considerations when placing national planning systems and cultures in their national economic development or welfare system contexts. Doing so helps unveil some of the parallelism or synchronicity in the development of seemingly different urban planning systems and cultures that might otherwise be overlooked. We obtain a sense of some the variety of 'national' developmental models and the way in which these constitute an increasingly polycentric 'global' geopolitical order (Bhan et al., 2018) – one with competing centres of urban planning imagination influence.

Comparative analysis can connect planning to other substantive concerns, such as housing (Kemeny, 1994),[7] and to contextual shapers, such as business systems (Whitley, 2007) and property markets (Berry and McGreal, 1995).[8] These particular systems might be subsumed within more encompassing national economic-development or welfare-system models. These models provide a sense of a wider world of urban planning than is typically entertained in much of the comparative planning literature. I summarize my thinking in table 6.1 above, which provides a partial and imperfect glimpse of urban planning's global variety.

Urban planning thought and practice are sufficiently long-standing, broad and internationally connected for any one nation to contain within it competing ideas and sympathies that can at points in history converge with those of other countries that might seem legally, administratively, culturally and geographically worlds apart. The idea of 'total planning' by a planning super-ministry (mooted after the war) in the UK came close to what was put into practice in countries across the Soviet bloc, and continued to attract interest for a while through exchanges of planning practitioners (Cook et al., 2014). Some of the theoretically ideal solutions of the nationalization of land and arrangements for compensation and betterment discussed at the same time ended up being deployed in large measure in Singapore. Model employment and management practices conceived in the US and Europe but difficult to implement there found their way into what subsequently came to be regarded as a distinctly Japanese manufacturing system (Dore, 1990).

Liberal markets

Much of the extant urban planning literature has emanated from and speaks to the liberal market economies of the UK, US and Australia and the social welfare economies of mainland Europe. I discuss these models only briefly here, but their openness ensures they continue to absorb and act as relays for ideas and practices internationally (Phelps and Tarazona Vento, 2015).

These national planning systems share some notable traits. Despite the label 'liberal market', urban planning *rules* have been an important part of city development (Talen, 2011), and each national system has had periods in which statutory planning, as part of broader 'social contracts' or 'deals', enjoyed a measure of influence.

Nevertheless, they are fundamentally mixed economies in which the private ownership of land and property and the right to transact them freely have rarely been impeded or strongly directed ideologically. The prestige of statutory urban planning has waned somewhat in liberal market nations: 'with no one and nothing in charge of maintaining an overall idea about what cities should be like, rules become the default overseers of urban form' (Talen, 2011: 5). As such, there is more than a sense of urban planning outcomes in these nations being a reflection of the distribution of power in society. Even so, contrasts exist across these liberal market economies in the overall direction of regulation (increasing public regulation of land-use rights in the US, elements of deregulation in the UK), and in how land use and its regulation reflect social stratification and its evolution: planning in the UK reflects long-established class divides (Healey et al., 1988), zoning in the US enshrines notions of social transition (Perin, 1977), while permissive planning of suburbs reflects socio-economic homogeneity in Australia (Gleeson, 2006).

It might be said that these economies' planning systems have been anti-urban in orientation. The bourgeois retreat from the city (Fishman, 1987) first observed in the UK and greatly encouraged in the US was perfected in Australia as the first suburban nation (Freestone et al., 2018). And despite the value of moving discussion away from these liberal markets, they retain value as reference points for some of the key trends that will probably condition urban planning more generally. Australian cities such as Melbourne, Sydney and Perth are perhaps *the* reference points for our suburban future. Cities such as Miami and Los Angeles in the US have their contributions to make to discussions of planning's response to superdiversity, while others speak to the sorts of urban decline produced from industrial overspecialization. The halving of Detroit's population in recent decades (Smith, 2017) is just the most extreme of many cases of urban stagnation which represent planning challenges that are not the mirror image of the challenges of growth (Galster, 2019).

European welfare models

In international context, the European social model is characterized by a high level of social protection and a more than residual emphasis on socio-spatial cohesion. Hence in table 6.1 we contrast the varieties of urban planning systems found across much of mainland Europe under this generic social model with that found in the UK and the mixed outcomes found in Eastern and East–Central Europe. However, there are a number of specific varieties that have been developed in the literature flowing from Esping-Anderson's (1990) original social-democratic, liberal and conservative welfare models (Nadin and Stead, 2008). Not least here is the salience of the family unit as a source of welfare found in Southern European nations (see Dunford, 1996), which raises thoughts of connections to some East Asian developmental states where the same reliance on family units can be found. 'Contrary to the ideal world of welfare states, the real world exhibits hybrid forms in every country' (Nadin and Stead, 2008).

In order to facilitate comparison at a global rather than European scale, I lump together the majority of mainland Western European countries in table 6.1. While it is difficult to sustain the argument that there is a single European welfare model, there is some sense from a global perspective in treating it as such due to some increasingly common elements therein (Nadin and Stead, 2008: 40). Urban planning across much of mainland Europe contains a stronger sense of the validity of state interventions (regulatory 'takings') in the public interest or for public works and is less localized than liberal market systems.

Enforcement varies even within OECD countries, of which mainland Europe provides a majority of members. While standards there are generally high and have been rising over time, enforcement in eleven of the thirty-two OECD countries surveyed was rated middling to low (OECD, 2017a). This feature of planning systems will be lower still across the many other global south countries, and especially in those models characterized by predatory practices in table 6.1.

Predatory systems

It is invidious to apply labels to peoples and countries, not least because these tend to stick unfairly. The betterment attached to land and property ownership (chapter 4) ensures that predation casts a shadow over urban planning, and a brief counterpoint to the emphasis on predation in the global south presented below is warranted. Fairfax County in Virginia is one of the wealthiest suburbs of the US, but as recently as the 1960s elected politicians were imprisoned over planning irregularities (Phelps, 2012b). As I write, the burgeoning outer suburban City of Casey has been put into administration by the state of Victoria, Australia, as a result of investigations surrounding the rezoning of land.

Many of the countries that might be considered to have 'predatory' urban planning systems are those with a reliance on natural resources. The 'resource curse' has often blighted the development of cities, not least because nation states have themselves struggled to secure authority and a monopoly of violence over territories pock-marked by enclaves of economic activity whose returns are the source of much political and paramilitary contestation. The regulatory and enforcement powers of planning are often weak in these countries, sometimes as a result of corruption in political and administrative systems in which planning permission is for sale. As a result, land-use planning follows actual development patterns rather than guiding them in advance and has few powers of enforcement; these are further prone to political influence and personal patronage. At the same time, planning powers for extracting betterment are least developed in these countries, where rampant private gain clearly poses significant collective action problems with regard to housing standards, infrastructure and environment.

Sub-Saharan Africa is a context that is difficult to generalize about, not least given debate about levels and trends of urbanization found across the continent (Potts, 2012). However for Ernstson et al. (2014) there are some unifying features that bear upon urban planning, notably a 'common sense of crisis affecting African cities', itself a product of the shared experience of colonialism, late decolonization and incorporation into the world system; the centrality and high visibility of informality; and the fundamental or partial senses of exclusion from the city by majority races and ethnicities. To this list we can add the way in which the urban planning imagination is caught up in questions of how urbanization has gone hand in hand with failed elements of democratization and decentralization (Olowu, 2018).

In African countries, 'planning and planners have absolutely failed to control urban space anywhere' and 'the overt disregard for the public good and sustainable urban practice' is the leitmotif even today (Parnell, 2018: 288, 289). However, 'coming to terms with why planning is so dysfunctional across Africa demands more ... than the simplistic rejection of modernism and discrediting of colonialism'

(Parnell, 2018: 289). In Africa, then, urban planning is not so much wrong as incomplete; incomplete geographically (as signalled in the sorts of control over space exerted by modern nation states) and historically (in terms of the lack of urban traditions relative to other parts of the world and the fundamental sense of a history of the city and urban planning largely waiting to be written). National systems of urban planning in African nations never covered the whole of national territories as they have come to do elsewhere in terms of cadastral records or complete coverage of land-use plans, however imperfectly designed and enforced. In part, this was a product of the comparatively late and extremely rapid scramble for Africa by seven imperial powers ensuing from the Berlin Treaty of 1885. In the colonial exploitation of Africa, enclaves of private enterprises abdicated responsibilities to the host environments in which they operated (Latham, 2001).

Colonial administration provided a remarkable opportunity to promulgate both philosophical tenets from home and practical experiences from elsewhere in empires.[9] Zoning regulations that supported intra-urban racial segregation and differences in density of development and in the legalization of informal areas (Silva, 2015) have been discontinued post-independence, but these dual mandate-related interventions have had lasting practical effects on both urban form and urban planning. Technical assistance and a development-planning paradigm predicated on Rostow's (1960) stages of growth towards an end state epitomized by American modernity filled much of the vacuum left by colonial powers, and found a receptive audience among many political leaders across Africa; however, they have also been part of the failure to generate indigenous urban planning traditions and capacity. Indeed, 'the foundations for the ... modernization paradigm of development had already been laid by the late colonial ... interest in Africa as a living laboratory for the application of science for the purposes of government and socio-economic improvement' (Odendaal et al., 2015: 290). Development-planning practice continued to have little sensitivity to local contexts, and the hope must be that African traditions – so often considered responsible for underdevelopment – will be recognized as a source of strategies for development and urbanization going forward (Njoh, 2006).

African nations may be emblematic of some of the challenges facing many urban planning systems in the global south, where in increasingly large portions of the developing world a situation is arising where some basic services are extended to informal areas but the underlying tenure issues are not addressed, making such investments accrue to intermediaries that control these areas beyond the regulatory purview of the state and on very exploitative terms for the urban poor. (Pieterse, 2008: 45)

Those local contexts continue to represent some of the more complex found globally, since they contain mixtures of statutory, customary and religious land tenure systems (Pieterse, 2008: 44). Outcomes cannot be read off from these different tenures, since chiefs who are central to the manner in which land becomes absorbed into the formalized peri-urban or suburban expanses of cities have been responsible for both progressive and regressive outcomes. The transformation of customary land holdings into statutory ones has further implications for the gendered nature of land ownership and livelihoods. Local government urban planning has taken shape, then, as a largely powerless activity within the local government 'shell' organizations of national states (Parnell, 2018: 292). Cities continue to be built out, by numerous individuals and by private corporations catering to aspirational club urbanism (Murray, 2017). In between, the formal system of national and local urban planning struggles to play a meaningful role.

The acronym KKN – *korupsi, kolusi, nepotisme* (corruption, collusion, nepotism) – that circulates in Indonesia summarizes the sorts of pressures that play into the urban planning system. These ingredients were highly centralized under the Suharto dictatorship (1968–98) when favours to individuals and businesses were doled out. Much of the urban sprawl of outer Jakarta – the new industrial estates and new towns in the likes of Bekasi and Tangerang – were licensed by central government. With the fall of Suharto and the decentralization of powers to the many poorly resourced local governments across Indonesia in the late 1990s, these KKN practices became decentralized in a period that might be characterized as little short of chaos, with local governments issuing many new regulations to raise taxes. Two decades on, much of the excess of this decentralized KKN has receded, partly as a result of the activities of international organizations which have sought to improve administrative efficiency and transparency.

The notable feature here is that the state – or state functionaries, including urban planners – may be a predator extracting rent, where the regulation associated with urban planning is a potential source of revenue. It is these countries that are the target of the World Bank's initiatives on the costs of doing business and on transparency. Institutional reforms have seen reductions in the numbers of ministerial or departmental approvals for businesses, and the creation of 'one-stop-shop' service centres in local governments – though there is evidence to suggest that there may be one shop but many windows as an instance of governments attempting to look like a state (Pritchett et al., 2013).

Dependencia (or dependency theory) is a powerful theoretical formulation developed to speak to the particular problems experienced by Latin American states (Cardoso and Faletto, 1979). Some specific elements of that formulation, such as the connection between foreign ownership of capital and lack of local economic development, have been discredited. Yet it is a perspective that continues to hold enormous popular and some analytical appeal and no little relevance to problems of urban, regional and national planning in these countries. Again, the state is weak and unstable economically and politically, given the reliance of political elites on a narrow base of export of commodities. There may be less short-term volatility in government and policy here compared to other predatory state forms. However, the dependence of political elites on resource extraction is often associated with extreme inequality and poor-quality urban environments, especially in resource-rich cities and regions, and has produced national revolutions and political coups with severe implications for not only the capacities of local governments and their urban planning efforts, but also the accumulation of citizens' positive experiences of participation in urban planning processes.

South Asian democracies

The South Asian nations such as India, Bangladesh, Pakistan and Sri Lanka have been labelled as ones in which processes of land development typically act to erase the voices of marginalized or subaltern populations (Roy, 2009b), but also call forth a wide variety of citizen practices resistant to the activities of states and clubs alike. These practices lend cities the appearance of being 'unplannable' according to the norms of land-use zoning and integrated infrastructure provision typical of received global north urban planning wisdom.

The use of legal means and formal urban planning is undoubtedly part of major land grabs by real-estate interests, repeated in many high-technology-industry and elite housing-oriented developments across a country like India (Goldman, 2011), from which those with insecure or unclear claims to land have been removed. However, land grabs represent a complex set of cases in which something of the powerlessness of the local government planner and the formality of local plans is revealed when set against elite political and big-business interests and the unruliness of some of the poorest in society. Organized occupations subvert formal planning processes increasingly attuned to the interests of urban property-development elites, and successfully stake claims to land, infrastructure and services, and even to a share of the betterment attached to major development or redevelopment of land (Benjamin, 2008).

South Asian countries have emerged powerfully as sites in which to observe elements of alternative, 'southern', urban planning practices as ones centred on squatting, repair and consolidation (Bhan, 2019). For Bhan, squatting, as a mode of occupying and laying claim to urban space, is an effective weapon of the weak – one born out of the sorts of substantive needs discussed in chapter 3. Significantly, it employs an imagination that embraces the uncertainties and multiple temporalities of statutory urban planning processes discussed in chapter 2. Repair is foundational to the incrementalism and self-constructed nature of citizen-led urban planning, and although 'chronologically subsequent' to construction (Bhan, 2019: 8), it nevertheless speaks to the phenomenological primacy of making use of 'things to hand' as part of our being in the world. It hardly needs saying that repair (along with reuse and recycle) is part of an ethos and set of practices that relate strongly to the possibilities for environmental and economic sustainability discussed in chapter 3. Significantly, 'a housing policy built on repair has an entirely different imagination of practice' (Bhan, 2019: 9), with financial models, capacities, institutional designs, standards and norms, and delivery models different from those mobilized by states and clubs. Consolidation speaks to the need to work with the imperfect realities of multiple means of service delivery that are in the process of establishment across cities. Consolidation emphasizes improving the coordination and governance of multiple socio-technical systems, rather than technical improvements to systems in part or in whole.

Developmental systems

The developmental states of East Asia represent a distinct but varied set of country approaches to urban planning. The legitimacy and powers of planning are strong given that governments are typically insulated from civil society, in some instances as a result of military dictatorship or de facto single-party rule. In these contexts, urban planning sits alongside evolving longer-term and comprehensive economic, industrial or sectoral planning in a way it rarely does elsewhere, though it may be fair to say that urban planning is consequent upon planning in these other spheres.

Developmentalism has been seen to be characterized by a synergistic connection between the public system and the private market system, where outputs of each become inputs for the other and

> the advantages of markets (decentralisation, rivalry, diversity and multiple experiments) have been combined with the advantages of partially insulating producers from the instabilities of free markets and of stimulating investment in certain industries selected by government as important for the economy's future growth. This combination has improved upon the results of free markets. (Wade, 1990: 5)

Wade draws a number of policy lessons from this: the need for national policies to promote industry investment within national territories; the potential for protection to help raise the competitiveness of domestic industries; the desirability of export-oriented (not import substitution) policies; a welcoming stance towards foreign MNEs, as long as they export intensively; the potential benefits of government influence on a bank-led financial system; the need to carry out trade and financial liberalization gradually rather than abruptly; the importance of establishing the primacy of economic development within national policy and bureaucratic hierarchies; the value of developing political authority before democratization to national economic development; and the value of developing corporatist institutions to national economic development.

The developmental states are characterized by the overriding substantive concern with economic development rather than, for example, environment protection and sustainability. Each of the main examples – Japan, South Korea, Taiwan and Singapore – represents variations on this theme. By placing economic objectives above others and positioning its Economic Development Board prominently within the ministerial architecture, Singapore rose in the space of fifty years to become a major regional corporate command-and-control centre. The rapid economic development of each of the main countries exemplifying developmentalism has been celebrated at one time or another, though each is now said to be grappling with phases of post-developmentalism.

Yet some notable differences have existed across these developmental states and are relevant for understanding urban planning and its effects. The complete and near-complete state ownership of land in China and Singapore contrast with the extensive private land and property ownership and transacting found in Japan, South Korea and Taiwan, and result in visible differences in the grain of urban development, including plot sizes and street patterns. Nevertheless, even in the latter three examples of the developmental model, the state exercises significant direct and indirect control over the urban distribution of business and government activities that shape the function and form of cities.

Some, though by no means all, of the developmental states alert us to urban planning's need to engage increasingly in the future with issues of population ageing, stagnation and decline (Ohashi and Phelps, 2020). Declines in birth rates have not been offset by immigration in the likes of Japan and South Korea. Here, local planning capacities may come to depend on innovative tax redistribution schemes, such as Japan's 'home town' tax which allows a portion of tax to be remitted to one's home town rather than one's place of residence.

China

While much urban planning in China is planning for growth (Wu, 2015), it is not growth or the build-out of cities per se that distinguishes urban planning in China. Historian Skinner (1977) has interpreted China as a series of regional urban systems. Thus, just as a national system such as the US is hardly one urban planning system at all, so too we need to rethink Chinese urban planning as representing more than one system despite the unitary basis of this nation state. There are a number of further reasons for alighting on the distinctiveness of China, it being best not to overstate either the imitative credentials of Chinese urbanism or the origins in China of 'Chinese models'.

First, China is the only continuing urban civilization to have survived from ancient times (Morris, 1994). It has a unique longevity that should not be discounted, however much surface appearances suggest otherwise. A civilization of this scale and duration 'perforce develops out of its own resources, tradition, and civilized genius' (Friedmann, 2005: xvii). Its longevity may owe much to the fundamental *pragmatism* of its people and administration. Moreover, in a history in which it is almost impossible to trace origin points, it seems likely that many instances of imitation may well be copies of copies from China. The Chinese urban planning system might be thought to have some of the variety found in most or even *all* of the systems in table 6.1. Parts of China, such as Guangzhou and Harbin, have something of the 'wild west' of liberal market economies and the overt corruption found in predatory states. Longer-term hopes are pinned on limited experiments with alternative – developmental-state – models gleaned from Singapore. The historical continuity and geographical complexity of China mean that its urban planning *encompasses* other systems.

Second, and in some contrast to the enduring character of its civilization, is the extreme speed and mixing of elements occurring in China today. China represents the display and projection of 'Shenzhen speed'. This speed is not an absence of planning; it is urban planning unparalleled anywhere at any time in history. It is clear that this sort of urban planning speed reflects some of the intense desires that can drive city making. Shenzhen may have no culture, but an absence of culture may be considered a culture of sorts too (Mars, 2008).

Third, it is problematic to categorize China as a nation state. Its current national borders are those of multiple imperial dynasties in which internal difference was as much a feature as administrative coherence. China's system of cities has been thoroughly administered for thousands of years – well before any such administration developed in Europe with the creation of nation states and capital cities (Rowe, 2013: 324). As administrative centres of empires, Chinese cities were never corporations as they were in Europe (Friedmann, 2005: 95); local administration of urban territories is something ostensibly copied from China (Stretton, 1975: 232).

Fourth, the history of the Chinese diaspora is ancient and so vast geographically as to present a cultural model unlike any other: we are dealing not so much with many 'Chinese variations on common western themes' as with global Chinese urbanism with its variations. The vast progeny of Chinese urbanity may be hard for us to comprehend. It is signalled in the ethnic Chinese diaspora, which reframes the process of globalization as one through which Chinese civilization has become transnationalized – a story of China 'unbound' (Friedmann, 2005: xvi). If the Chinese diaspora is the informal means of this unbounding of China, the Belt and Road Initiative (BRI) is, of course, its formal counterpart.

Born globals

A set of cities have been 'born global' and often dominate symbolic-ally or materially in terms of economic activity or population in their respective 'national' planning systems. It is important to observe them closely and draw planning wisdom from them. They have a specificity derived from the unique geopolitical and geo-economic niches open to small or island states (Baldacchino, 2010) and in a peculiar, narrowly focused fusion of club and state urban planning imaginations. Worryingly, despite inimitable and perhaps undesirable aspects, they have increasingly become super-ficial reference points for politicians and urban planners. Indonesia's planned new capital city on Kalimantan may be just the latest example of an attempt to meld every possible planning cliché. The seductive powers of born-global cities as urban forms may yet draw politicians and populations to them in ways that urban planners have otherwise been unable to.

There is never a tabula rasa upon which settlement takes place. However, born-global planning systems have developed on 'clean slates' by and large. To the extent to which we can reasonably identify these urban planning systems, we might say that our story begins with Singapore. At independence Singapore was a tiny island with no natural resources and no economy to speak of. Today it is firmly on the map of world cities of note. Urban planning was absolutely central to this achievement, and the absorption, adaptation and further refinement of practices originally imported has quietly enabled the city-state authorities to become international leaders in urban planning thought and practice (Miao and Phelps, 2019).

Shenzhen in China is another early exemplar of born-global urban planning, having grown from a fishing village of 13,000 at the time of

Deng Xiao Ping's policy to open China from the 1980s to a metropolis of over 10 million today. 'Shenzhen speed' is simply the most extreme instance of, and in large measure the model for, the rapid build-out of cities across China. Yet for all its speed, and surely because of its embeddedness in an ancient urban system, Shenzhen has not forced itself into the popular, political and policy consciousness as much as Singapore and Dubai.

Cities such as Dubai in the United Arab Emirates and Nur-Sultan (formerly Astana) in Kazakhstan have appeared as instant cities that stand for global aspirations. They presage planning of a distinctly different order and for which surely very few historical antecedents can be found. From the very outset these cities have been conceived to consciously act as articulators in the networks and flows of global urban economic connections. Dubai seems to have escaped the indifference of consumers and much of the ridicule among commentators reserved for Nur-Sultan (formerly Astana), which is sometimes represented as the worst example of the pick-and-mix planning that I warn against in the concluding chapter.

Dubai is uniquely planned as a new staging post to capture economic possibilities as the gravity of the world economy shifts back to the east (Elsheshtawy, 2009). It has been described as a corporate city (Kanna, 2011) and could be considered to fuse the categories of club and state planning. It is unique in housing 200 nationalities in a city with little if any historic core or native population (Elsheshtawy, 2009). Dubai is a prime example of the seductive power of a city designed with the international consumer in mind (Acuto, 2010; Kanna, 2011). Moreover, it is a city designed specifically to cater to the proliferating aspirations and desires I consider further in the next chapter. If Dubai can draw on a vast tourist market, it may nevertheless struggle to seduce beyond the initial experience. Its visual fascination may quickly be exhausted under the tourist gaze and be in need of near-constant re-production and re-enchantment. Questions therefore remain over its economic sustainability, and extend to the maintenance and management of both extensive new infrastructures and the mix of populations it caters to.

The club properties of these born-global planning systems ensure that the urban populations concerned are self-selecting. Nevertheless, powerful technologies of societal management and ideological manipulation may be among their legacies in the sorts of international policy exchange which I consider in the next chapter.

Conclusion

National specificities inhere in the legal and administrative bases of urban planning and have been overlaid by local practices. These provide some basis for the distinctiveness of urban planning imaginations found internationally. Proof of the value of such contrasts comes from the Singapore government's careful curation of the urban planning ideas, policies, practices and tools – the 'wisdom' – which brought it 'from third world to first', while other nations appear content to forget lessons from their own planning past. Yet, equally, the distinctiveness of all national and local planning systems and cultures may be under threat if urban planning becomes reduced to the sorts of pick-and-mix planning that I noted in chapter 2 and return to in chapter 8, where I consider the prospects for urban planning.

Then again, the commonalities that unite countries across continental regions and across national economic development or welfare models present new opportunities for the exchange and development of urban planning theory and of learning regarding urban planning policy and practice. Among these countries and localities are those experiencing strong population growth, including through large-scale and diverse immigration and/or rates of urbanization, on the one hand, and those experiencing population stagnation and shrinkage, on the other hand. The two processes are not mirror images of each other (Galster, 2019) but seem set to unite citizens, clubs and states and their preoccupations and imaginaries in otherwise very different national, continental and economic-development model contexts.

7 Exchanges: what are the global connections in planning?

Introduction

There are different urban planning systems and cultures, but powerful forces of convergence in planning thought and practice have also been apparent. Trade has long facilitated patterns and processes of cultural exchange (Bayly, 2000) including those focused on cities and the planning of cities. Indeed, 'The development of modern planning practice from the late nineteenth century was, from the outset, an international activity' (Ward, 2005: 119).

The geohistorical story I present of the international exchange of urban planning ideas is not one of *either* difference *or* commonalities, but one of their coexistence in variations on common themes. Saunier (2002: 522), commenting on the international exchange of urban planning ideas in the early 1900s, notes: 'connections ... can exist, and differences can be discerned and acknowledged, without preventing the development of a universalist discourse whose very terminology may not always be universally understood or accepted'. As Friedmann (2006: 228) describes it:

> It could be argued that the current era of globalization and the insertion of cities into the 'space of flows' of global finance, information, and cultural exchanges will eventually lead to a greater homogenization of practices ... But even if this were the case on the technical side, actual planning practices must still respond to the particular conditions under which they operate, conform to the prevailing political culture, accommodate to its institutional settings, adapt to the limitations of resources for local development, battle with entrenched interests and traditions, and so gradually evolve its own national and even local style.

Taken together, this and the preceding chapter are designed to render a relational geography of urban planning as rooted in and addressed to distinct places (neighbourhoods, cities, nations) and at the same time exposed to exchanges that transcend those places.

The previous chapter emphasized the meso-historical inertia inherent in the organization of urban planning into national planning systems and cultures. This chapter emphasizes the variety of actors – citizens and clubs as well as states – and indeed the mixes of these actors that lie behind these seemingly stable contrasts between statutory planning systems and cultures. The 'histories of globalization have a complexity of networked action which means that few if any actors have the synoptic capacity to be rational' (Braithwaite and Drahos, 2000: 548). As such, and despite continued elements of the bureaucratization of world society including urban planning, the urban planning imagination continues to seduce and persuade us all.

The who, what and how of exchange

If we are to consider the international exchange of urban planning ideas and practices, then we need to recognize what has been exchanged, how such exchange has typically taken place and by which actors. Ideas from international relations and sociology prove useful as guides to the diversity of actors involved in exchanges of the urban planning imagination and the typical means involved in and elements of such exchange. States remain important actors in and orchestrators of international exchanges of planning thought and practice, but citizen and club actors emerge as important in their own right.

The international relations literature has generated an extensive list of potential processes of policy exchange underlying the gradual emergence of global governance. The different mechanisms or regulatory principles that Braithwaite and Drahos (2000) identify are: military coercion (i.e. the threat or use of military force); economic coercion (the threat or use of economic sanctions); systems of reward (valuing compliance); modelling (observational learning with a symbolic content); reciprocal adjustment (non-coerced negotiation and mutual adjustment); non-reciprocal exchange (movement towards common rules and approaches without all actors believing in them); and capacity building (the provision of technical assistance to fulfil a desire to meet real or perceived global standards). And it is important to realize that 'the globalization of regulation is never the product of one mechanism' (Braithwaite and Drahos, 2000: 542).

Military and economic coercion have been important in the spread of the global north planning imagination and associated substantive concerns, wisdom and methods. Much of the urban planning imagination used in the building and extension of cities secured coercively rested on the work of individual surveyors, architects and engineers in the service of states and their empires, but has since emerged within epistemic communities of practice at least partially detached from states.

Military and economic coercion cannot be consigned to the history books, but they have been replaced with those mechanisms that constitute what Boli (1999: 273) terms 'rational voluntarism' (modelling, reciprocal and non-reciprocal adjustment, and capacity building) and which have emerged strongly as important drivers of policy exchange. Modelling emerges – both historically and in the present – as the dominant mechanism promoting a measure of global business regulation (Braithwaite and Drahos, 2000: 250–1). Yet the process of modelling is itself heterogeneous, as Braithwaite and Drahos (2000) seek to capture in the categories of 'model missionaries' and 'model mercenaries' (composed of actors who are locally recruited), 'model mongers' (those, notably globally operative consultants, who are able to foist ready-made models on recipients due to satisficing behaviour), 'model misers' (where selectivity of adoption is related to a lack of resources to generate solutions from first principles) and 'model modernizers', each of which has its exemplars in the field of urban planning. Braithwaite and Drahos note that it is private sector actors, not states, that have been the most important in driving much global regulation in recent decades. Together with NGOs, then, private enterprises emerge powerfully as a large and diverse set of club actors in a more fully bureaucratized, global, interstate system promoting the exchange of urban planning thought and practice.

The sociology literature provides a different but overlapping classification of 'modes of power' as a means of understanding some of the exchange of ideas and practices between different urban planning systems and cultures. Allen (2005) has highlighted the geographical dimensions of several different modes of power, including domination, manipulation, seduction, authority, inducement, coercion and persuasion. Domination and coercion map more or less onto Braithwaite and Drahos' (2000) mechanisms of military and economic coercion, and inducement onto their systems of reward. Authority emerges as the mode of power most clearly exerted by way of the bureaucratization of urban planning within the international interstate system.

This leaves persuasion and seduction as perhaps under-appreciated ways in which urban planning ideas have travelled. Indeed, the seductive power of the city's image has ensured that these 'soft' forms of power have been important to the exchange of urban planning thought and practice stretching back to ancient times. Soft power is the ability to structure situations so that others develop preferences or define their interests in ways that are consistent with one's own, and draws significantly on the attraction of culture and ideology (Nye, 1990). Are these soft powers of the urban planning imagination once again to the fore? It is the seductive power of the urban planning imagination that I emphasize as a property driving international patterns and processes of exchange in urban planning theory and practice today, and one that provides a basis for reviving the reputation of professional urban planners alongside their more illustrious built-environment colleagues. Seduction is a power in the process of exchange that has seemingly been better mobilized by citizen and club actors than by states or statutory planners. States are hardly ignorant of the value of exercising soft power, but it is cities, their beauty and what they represent of civilization that travel, not the abstractions that are national identities and national states – a fact apparently better understood at present in different ways by citizen and club actors. The future of statutory urban planning imaginations rests more than ever at this juncture on its potential for emotional intelligence (Hoch, 2019) when seeking productive engagements with the seductive ideas propagated by citizens and clubs.

We should be wary of thinking that urban planning thought and practices travel unchanged. Of necessity, ideas and practices and even 'models' are adapted to local circumstances and are therefore transformed in the process of exchange. We might speculate that it is the malleability of key ideas that has ensured their popularity internationally. Moreover, the visibility and transferability of different elements of planning policies vary. It is the techniques (know-how, operating rules, methods) associated with planning policies (programmes, projects, practitioners) rather than the ideas (principles for action, philosophy) and the wider organizational culture and setting that are most transferable (Stead, 2012). Design may be 'whole-city big' (Molotch, 2004: 212). However, as Osterhammel describes of global transformations in the nineteenth century, 'model cities and architectural styles interacted with each other in different ways. The latter could be more easily copied than the former, but the cultural "spirit" of a city almost not at all' (Osterhammel, 2015: 312). Therein lies much of the confusion and frustration in the quest to make better urban places.

Nation states, empires and urban planning

The historical relationship between cities, nations, empires and states is a complex one (Curtis, 2016) and the urban planning legacies of these connections in Southeast Asia, for example, could hardly be described as entirely bad, given the ordering of spaces to mitigate externalities, including public spaces and amenities and serviceable water and sewerage infrastructure that remains. Elsewhere, as in Africa, the colonial legacy might be better described as a wound, generated by the hastier and greedier land grabs of states and clubs in the form of private enterprises. Thus, important accounts of the international spread of urban planning ideas in the modern era have emphasized the imposition of urban planning thinking and practice on colonies, or else the borrowing from imperial powers (Ward, 2010). One lesson to be drawn from empires ancient and modern is that it was often easier to impose urban order away from the home of empire itself. Morris (1994) contrasts the relative chaos of Rome with the many new grid-iron military camps established during the expansion of the empire; Romanization was achieved through urban-ization and an ordering of the urban overseas. Similarly, modern imperial powers were often better able to experiment with urban ideas and practices overseas – not least due to treating colonies as a *tabula rasa* or *terra nullius*, in a manner impossible at home.

Ward's (2010) discussion highlights both how more than one mechanism or mode of power was always implicated in the urban planning of empires (since imposition relied on authority and coercion) and the demand for some of the urban planning offered by imperial powers (in that borrowing involved elements of synthesis, selection and uncritical reception). It is hard to overestimate the development of urban planning thought and practice as it was played out in the overseas colonies of empires: 'in some cases, especially in the French and Japanese empires, colonies served as relatively unconstrained legal test beds for planning powers that were only later adopted at the heart of empire' (Ward, 2005: 129). India, as the 'jewel in the imperial crown', proved an important but imperfect test bed for British administrators and urban planners. These included Patrick Geddes, the vast part of whose practical works were undertaken there (Meller, 2005). Ebenezer Howard's Garden City ideas found favour in the planning of New Delhi (United Nations Human Settlements Programme, 2009: 51). Elements of today's urban planning in the UK reached their zenith at the edge of empire in Adelaide, Australia, in 1836–7 (Home, 2013). Here the elements evident were: (1) a policy of deliberate urbanization; (2) allocation of land rights; (3) a plan drawn up in advance of settlement; (4) wide streets in geometric order; (5) provision of public squares; (6) standard plot sizes; (7) reservation of plots for public purposes; and (8) a physical distinction between town and country with a belt of common or green land. Fully one hundred years ahead of the introduction of the London green belt, one of the most celebrated and defining features of the UK planning system emerges as a policy pioneered abroad. This is no accident, as the legal fiction of *terra nullius* (Jackson et al., 2017) ensured a land grab of spectacular and brutal proportions. This was urban planning in the stripped-down form of surveying; the parcelling of land ready for the maximum exploitation of the private ownership of land and property.

Paradoxically, the expansion of empires was a competitive process in which little regard was given to experiences in other empires (Home, 2013: 54). In this respect, rather than promoting a measure of convergence, the projection of imperial power abroad onto overseas societies helped crystallize national planning systems, traditions and even specific planning ideas, policies, tools and practices *at home* in European cities. Moreover, while it is important to provincialize European experiences in a global history of urban planning, the urban planning of former empires is not merely of historical interest, not least because 'the built heritage and cultural legacy of former colonial empires ... remains relevant to the negotiation of postcolonial identities' (Home, 2018: 102). Indeed, the considerable inertia that inheres in planning systems and cultures and within cities as built artefacts ensures that more than physical traces remain. Thus, even today, in many parts of the world, planning systems are in place that have been imposed or borrowed from elsewhere. In some cases, these 'foreign' ideas have not changed significantly since the time they were imported. Planning systems and urban forms are inevitably based on particular assumptions about the time and place for which they were designed; but these assumptions often do not hold in other parts of the world and thus these systems and ideas are often inappropriate in the context to which they have been transplanted. (United Nations Human Settlements Programme, 2009: 50)

Among European nations, an increasingly professionalized class of urban planners (architects, surveyors and engineers at this time) were part of a larger 'fund of imperial knowledge' that was 'selectively applied to the development of the rapidly growing system of increasingly linked port cities and towns worldwide ' (King, 2004: 84).

The visible traces of this fund of knowledge can still be seen in the quarters and concessions found in cities in the former colonies of European imperial powers. Neighbourhoods took on important legal dimensions which further ensured their economic distinctiveness and value to imperial powers. Shanghai's Bund represented an export processing zone (EPZ) of its day: a riverside strip of land demarcated as extra-territorial for the purposes of expediting trade (Taylor, 2002). The port of Tianjin in China was carved into nine distinct areas from the late 1800s by the British, Italians, Germans, Japanese, Belgians and French, among others. Commonly, particular quarters or concessions of overseas cities would be planned and settled by expatriates of imperial powers. Even without additional legal restrictions on land use, these neighbourhoods would have had, and often retain, their preferential status as occupying prime locations with distinct road patterns, architecture and green spaces. Their distinctiveness continues to make for exclusive residential neighbourhoods or vibrant commercial and entertainment districts within rapidly changing cities (as in the former French concession of Shanghai). Elsewhere, limited remnants of European architecture form the basis for present-day urban planners to leverage cultural or entertainment districts, as in Tianjin's Italian quarter. Less visibly, the infrastructures of former colonial powers continue to have effects today. The Dutch sewerage system of the port of Pekalongan has helped ensure the continued inner-city presence of polluting industries such as batik, in contrast to the exit of the industry from the ancient sultanate city of Surakarta. To Ginkel and Marcotullio (2005: 17), the water and sanitation systems of colonial powers now represent something of an infrastructure time bomb in some Asian nations, since they are overloaded and not extended in the way a typical city network might be.

Moreover, principles of 'indirect rule' and 'dual mandate' (securing imperial economic opportunity while preserving local traditions), which became generalized across the main European empires (Home, 2018: 94), played into traditions of segregation found in the urban planning practised in colonial cities. Here different substantive concerns of urban planning – land use and housing, health and sanitation – combined in powerful discourses regarding the need for urban segregation (Home, 2018: 95). Some European powers mobilized sources of labour from different parts of their empires, with concessions or neighbourhoods of colonial cities taking on distinct racial or ethnic complexions. In one of the many ironies that abound in the geohistory of urban planning ideas and practices and their unfolding effects, the distinct ethnic Chinese–Malay–Indonesian–Indian mix of Singapore is now mobilized in the promotion of the island as a unique laboratory for health research. While the colonial cities of European imperial powers were hardly unique in having ethnically segregated neighbourhoods, the difference was that this became a matter of urban policy and planning (Home, 2013: 124).

By the 1800s, 'ideas and political movements jumped across oceans and borders from country to country' (Bayly, 2000: 3). Urban planning professionals were important in the building of colonial cities, but these cities more than most embodied 'plural societies, multiple identities and "intertwined histories"' (Nasr and Volait, 2003: xxi). In the construction of colonial cities we can observe the production of a 'complexity in urbanistic exchanges' in which it was often unclear what was local and what was foreign (Nasr and Volait, 2003: xv, xiii–xiv). Indeed, the fashioning of some measure of imperial hybridity seems to have been foremost in the minds of architects and planners in the colonies. In its planning it was argued that New Delhi was to be distinctly imperial, rather than of any existing place (Home, 2013: 152).

During the late 1800s and early 1900s, surveyors, planners, architects and civil engineers formed an international movement of sorts (Osterhammel, 2015; Saunier, 2001). This was 'the only moment when ideas on the future of the life in large cities were discussed by international bodies unfettered, to a large extent, by the constraints of practical politics' (Meller quoted in Cherry, 1996: 34). The urban planning implicated first in the creation of European nation states and subsequently in the expansion and maintenance of their empires culminated in an 'Atlantic era' (Rodgers, 1998) of public policy exchange, in which the city came to the fore as the object of concern and territory of action (Saunier, 2001: 382). Urban policy exchange in this era produced 'an environment where ways of judging, appre-hending and acting on the city were defined, where expertise and professional legitimacies were created, [and] where knowledge and disciplines were constructed' (Saunier, 2001: 382).

One legacy of this era has been that 'the West remained a spatial and ideological construct that existed only under American guarantee' (Ward, 2003: 102). This is ironic in that 'the US has never developed an urban planning tradition commensurate in influence with its broader global significance' (Ward, 2003: 102). The thought that China – a country whose citizens, club and state urban planning actors are widely held to imitate others – may be beginning to exert an effect globally through the likes of its Belt and Road Initiative (BRI) without such a coherent or influential home urban planning model is just as ironic. China's emergence as a major propagator of urban planning ideas and practices internationally raises the more general question of the geographies of any new era of policy exchange. What are the prospects, then, for an era marked by the global south–south or Asia–Pacific exchange or Sinicization of urban planning?

Urban planning's global iron cage?

As part of a global level of cultural and organizational formation and the creation of major international institutions after the Second World War, urban planning thought and practice began to take on a genuinely universalistic character. Correspondingly, 'a simple uprooting and transplanting transfer process could be justified by the belief in a single, "universally valid" pathway for human social development' (Healey, 2012: 191). The numbers of inter-national non-governmental and inter-governmental organizations had increased exponentially from 200 in 1900 to around 4,000 by 1980 (Boli and Thomas, 1999a), to the point where these organizations and professions began to propagate claims to universalistic urban planning knowledge. The circulation and interaction of the distinct and competing national urban planning knowledges of imperial powers have increasingly given way to the enactment of 'global cultural models' (Boli and Thomas, 1999a: 173). And for some, this is a world society

> made up of what may ... be called 'rationalized others': social elements such as the sciences and professions ... that give advice to nation states and other actors about their true and responsible natures, purposes, technologies, and so on. Rationalized others are now everywhere, in massive arrays of international associations ... and epistemic communities ... generating veritable rivers of universalistic scientific and professional discourse. (Meyer et al., 1997: 162)

While urban planning theory and methods are undergoing a structural shift in recognition of the reality of global south conditions and their possibilities, it is salutary to confront the thought that the formal bureaucratic urban planning world surface – of nation states, international inter-governmental and non-governmental organizations – is still in the process of forming. In this regard, the commodification of place and the urban planning associated with it are products of policy-making processes which have themselves become more bureaucratic (Majone, 1989).

The Europeanization of urban planning

Add the existence of the ancient empires of Greece and Rome to the breadth and depth of the modern empires established by European nations, and it may be little exaggeration to speak of the Europeanization of urban planning worldwide. In the modern era, eminent urban planners such as Daniel Burnham in the US were in awe of the urbanity of Europe, and the likes of German zoning swiftly made an impression in the US (Ben-Joseph, 2012; Hirt, 2007; Talen, 2011). It is only recently that the good life represented by the suburban single-family home model perfected in the US has begun to be exported in earnest (Beauregard, 2006).

The Europeanization of urban planning in the present is perhaps more narrowly confined to Europe than ever before. Spatial and land-use planning do not figure as an objective of any EU Treaty, to the point where the European Commission does not have competence in urban planning matters. Nevertheless, Europeanization can be conceptualized as 'a discursive and cognitive process whereby certain normative concepts and "spatial planning ideas" are encouraged by

EU institutions, programmes and documents such as the ESDP [European Spatial Development Perspective], and through debate among actors from across the EU but are then reshaped by local actors' (Dühr et al., 2010: 360).

Europeanization's influences have been several, at times subtle, but quite pervasive as a result of the depth to which regional economic, monetary and political integration has been pursued in this part of the world. Europeanization is a process that embodies elements of 'upload' (from localities), 'download' (top-down diffusion and dissemination) and 'cross-loading' (inter-locality exchange) of ideas and influences (Dühr et al., 2010). However, there remains a measure of debate regarding the real depth of such processes of exchange. For Kunzmann (2005: 236), 'Best practice transfer from one country to another as promoted within Europe by the European Commission is extremely superficial.' Part of the reason for this is surely the enormous variation in local government capacities and modes of operation that exist across Europe, and the language differences that can impact on the exchange of ideas (Phelps et al., 2002).

Yet for Faludi the exchange of urban planning thought and practice across the EU has been little short of a 'learning machine', to such a degree that 'despite the lack so far of any agreement on an EU role in planning, shared learning has brought planning to a point where a return to inward looking national planning is inconceivable' (Faludi, 2015: 263). The early emphasis in the work of EU institutions was on the shaping and deepening of its common market with the elimination of non-tariff barriers. To the extent that planning represents a non-tariff barrier, there has been some indirect – ideological and rhetorical – pressure driving a measure of convergence across the EU, since the bureaucracy of planning systems can restrict international flows of investment (Healey and Williams, 1993).

The ESDP produced in 1999 was the attempt to fashion a common approach to urban planning across the EU. The ESDP's influence is indirect, since it has no legal status: it is advisory only. It is a product of its time, notably in its reference to 'polycentricity', a word which has since receded from the planning vocabulary just as the term 'sustainability' has come to the fore. Thus,

> while, on the surface, certain spatial concepts (such as polycentricity or sustainable development) seem increasingly present in planning policies at all scales across EU countries, not least because of the European

circulation of 'planning ideas' encouraged by transnational cooperation programmes, such concepts are open to very diverse ... interpretations, and are reinterpreted and adapted by domestic actors for their own purposes. (Dühr et al., 2010: 373)

Other activities of the European Commission have had more direct, significant, consistent and far-reaching effects on national and local planning systems. Those policies that deliver substantial funding for infrastructures and projects have naturally exerted some influence on national approaches across the EU (ESPON, 2018). Other strong Europeanization effects are related to the transposition of EU directives into national laws. Among these must surely be a large number of environmental directives; those with the greatest bearing on urban planning relate to the likes of the abstraction, treatment and discharge of water and the methods of Environmental Impact Assessments (EIA) associated with development applications. These have continued to bite in the southeast of England, for example, checking even much-needed housing development where the quality of river and sea water stands to be adversely affected because treatment capacity and technologies have reached their limits.

Global urban planning governance?

While from today's vantage point, the limits of imposing or adopting alien urban planning imaginations, systems, policies and practices are increasingly recognized, it could be argued that the coercion of states and empires has been replaced by a more subtle and pervasive imposition of urban planning norms under the aegis of international inter-state organizations (such as the UN and OECD), local, national NGOs, professional associations and numerous networks of city planners. Something of this is captured in Rykwert's (2000: 142) lament that the buildings of institutions of global governance 'have settled cuckoo-like in various cities without making any contribution to them'.

On the one hand, the UN's SDGs appear to have re-enchanted a global 'marketplace' for urban planning ideas. It is a market into which any number of club actors – consultancies, NGOs, privatized arms of local and national governments – and citizens have been enrolled. The inclusion of a stand-alone urban SDG of safer, more inclusive, resilient and sustainable cities has strengthened the relationship between notions of sustainability and urban planning expertise (Owens and Cowell, 2011: 20), and is predicated on an understanding of global processes working laterally *through* cities rather than operating *upon* them as external forces (Barnett and Parnell, 2018). UN-Habitat has done much to publicize sustainability in city planning. The World Urban Forum organized by UN-Habitat in Vancouver in 2006 had 9,600 delegates from 100 countries (McCann, 2011: 119). Such policy arenas are 'the institutional sites where members of policy communities come together to develop ideas and actions for urban futures. They act as nodal points for stakeholders and are places where critical decisions are made' (United Nations Human Settlements Programme, 2009: 89).

On the other hand, these new international arenas of urban planning are often characterized by inequalities of access:

> First, the more 'open-access' arenas become segmented off from the main nodes for urban policy development. ... Second, the agenda is set by the public sector and the participants are 'the usual suspects' (i.e. the same group of people appearing in different arenas in slightly different combinations and compositions). Participation of any 'outside' stakeholders is marginal and policy is made, often behind closed doors, by a small, yet powerful, group of government officials and a few large businesses. Such marginalization of informal forums and their late inclusion in the process leads to a sense of democratic deficit and distance between governments and citizens in urban policy processes. (United Nations Human Settlements Programme, 2009: 90)

This globalization of the business of urban planning has been accompanied by a shift from comprehensive to project-based planning. The charge made is that 'the policy frameworks and tools promoted by mainstream development agencies such as UN-Habitat and the Cities Alliance do not help local actors to make decisions in a way that advances transformative urban change' (Pieterse, 2008: 35–6).

Cities and city systems were important independent and path-shaping variables in the formation of nation states in Europe. Tilly (1989) identifies state 'coercion intensive', city-centred 'capital intensive' and city-state 'capitalized coercion' paths to this formation. The contemporary urban age might, then, be regarded as a drawn-out moment in which the organizational and governing capabilities of cities have once again come to the fore. The competitive and diplomatic orientations of local governments – including those relating to urban planning – have become increasingly extraverted (confidently outward looking) or extra-jurisdictional (Lauermann, 2018). This is a reawakening of a logic that existed prior to and during the formation of nation states in Europe, when, for example, monarchs and states were concerned to gain total control of territories while city author-ities, in contrast, typically cared little about borders and sought to promote easy communication aligned with trade (Blockmans, 1989: 218). For Carolini (2015: 269), the soft power exerted by individual countries is 'especially clear on the urban front, where these countries leverage their own experiences and challenges with urban growth as a rationale for their diplomatic outreach on other countries with upcoming urban growth'. The increasing number and variety of international city networks now evident might be regarded as a new urban diplomacy practised autonomously or on behalf of nations by city authorities (Lauermann, 2018), in which there is a complex relationship between nation, state and city under contemporary globalization (Curtis, 2016). To be sure, some city networks are little more than vehicles for lobbying national and supranational bodies for finance and a favourable policy and regulatory environment. However, in some policy spheres – such as climate change – cities have effectively bypassed states (Curtis, 2016). Examples include the C40 Climate Leadership Group, International Council of Local Environmental Initiatives, United Cities and Local Governments, World Association of Major Metropolises, City Protocol Society and the networks funded by the Rockefeller Foundation. As we saw earlier, for some, these networks and the experiments they embed hold out the possibility of cities as laboratories for urban planning. Questions remain, however, over the scalability of experiments and in some instances the desirability and implications of club- and state-led laboratory 'experiments' upon citizens.

Before we get carried away with conspiracies of actually existing global governance, it is as well to ponder the limits of this emerging architecture. The UN-Habitat's Millennium Development Goals (now the SDGs) challenge of removing 100 million slum dwellers from poverty by 2015 seemed ambitious, until one realized that it implied that '90 percent ... of slum dwellers worldwide would have to accept that they [were] not part of this deal' (Pieterse, 2008: 113). Historically, it was the weak cohesion between cities within networks that resulted in their eclipse by modern nation states (Blockmans, 1989), and we should bear this in mind when interpreting the depth and permanence of city networks in the present. Urban planning still appears rooted in the nation-state system (Acuto and Rayner, 2016), while few nation states appear to have emulated or learned from wiser cities.

Clubs: planning for extraordinary desires

If culture is the most profound driver of globalization, then we need to consider the nature of its mobilizing power. Culture seduces in its appeal to the desires of individuals, clubs and states. Culture's power to seduce has long existed; it gathered strength in modernity and in a first global economy that had emerged by the late 1800s. Arguably, the seductive power of culture has become pervasive today, though this term rather underplays the way in which urban planning thought and practice now fuse all of the different actors – citizens, clubs and states – in ways which are profoundly functional to modern capitalism.

The seductive power of urban planning

Perhaps the best-known instance of the seductive power of urban planning is the copying of the Chinese city of Chang'An as the inspiration for the creation of Kyoto in Japan in 710, though the copying of city layouts was apparent in Harrapan culture in India thousands of years before (Morris, 1994: 405, 31). Recent copies exist, such as the Soviet Union's first major communist-system experiment with iron and steel production at the new town of Magnitogorsk in Siberia, copied from the existing town of Gary, Indiana (Kotkin, 1995). The story of the planning of Magnitogorsk and exchanges between British and Soviet planners in the early post-war years (Cook et al., 2014) underlines how the very idea of reconciling individual and collective needs at the heart of modern urban planning makes it a uniquely unifying activity, capable of transcending some of the biggest modern ideological, cultural, political and policy divides among nations.

The grid is perhaps the commonest urban form in history and this must surely be due to some intrinsic appeal to the worst of our nature. This is the seductive power of urban planning in the service of the pursuit of wealth. 'No better urban solution recommends itself as a standard solution for disparate sites, or as a means for the equal

distribution of land or the easy parcelling and selling of real estate'
(Kostof, 1991: 75). This is the urban plan as the vector of real-estate
appreciation. Despite its sometime shocking superimposition on
nature's curves – as in Melbourne's 'Hoddle Grid' in Australia – the
grid's appeal is to render the city a container to be filled, emptied
and refilled in an endless series of land and property transactions.
Melbourne's grid was an extreme act of theft, resting as it did on the
legal fiction of *terra nullius*.[1]

'As a practical ideal, Ebenezer Howard's *Garden Cities of Tomorrow*
(1902) conquered the world, including the United States' (Faludi,
2015: 268). Although the Garden City was widely adopted by many
states, it is worth remembering that it was a club initiative. It is
no accident that the idea had such appeal; it was intended by these
individuals and corporations as an act of persuasion, much as are
the gated community developments of the present. The malleability of the idea must be part of its appeal, since 'the popularity
of the Garden City as a principle of town planning was its extreme
flexibility; its relatively easy ensconcement into any ideology' (Kostof,
1991: 77). The Garden City and new town ideas first propagated in the
UK continue to represent an ideal in some contexts, have travelled
widely, have been instituted differently in different national settings
and have been regarded rather differently in each. Indeed, new towns
remained attractive to postcolonial nations seeking to develop new
capital cities such as Abuja in Nigeria, Dodoma in Tanzania and
Chandigarh in India (Home, 2018: 96).

'The basic notion of building a new town is simple to grasp and
is capable of capturing the imagination of the public and politicians.
This factor undoubtedly explains in large measure why the concept
has been so widely adopted' (Williams, 1986: 89). Moreover, some
of the work of empires in the projection of modern urban planning
thought and practice was done by multinational enterprises. The distil-
lation of emerging international practices was undertaken by these
enterprises in the carving out of enclave order from the surrounds of
overseas colonies. Mining towns, such as the nitrate mining city of
Chacabuco in the Atacama desert of Chile, combined ruthless exploitation of workers with model urban planning, including the rational
layout and accommodation of population with all of the facilities of a
town, such as a hotel, a theatre, a market and a central plaza. Despite
their remoteness and extreme climate, these towns and cities could
be globally connected in a way that some provincial towns in the

global north today are not. Planned towns such as Chacabuco might be thought an extravagance in an urban age of fly-in-fly-out mining labour forces, yet they provide a hint of another legacy of the state–club nexus found within empires that continues into the present: the use of urban space for conspicuous consumption and the maintenance of inequality (Home, 2013: 227). Indeed, we might speculate that it is the power of urban planning to seduce that has promoted a measure of alignment in the planning of states and clubs in the emergence of a transnational capitalist class (TCC) (Sklair, 2001).

Urban planning by and for a TCC

'Particular cities have distinctive planning styles – styles that often shape global forces as well as being shaped by them' (Newman and Thornley, 2011: 268). Key to this dialectical relationship between world cities and global urban planning practices is the last part of Friedmann and Wolff's (1982) world-city definition: that the elite of world cities increasingly constitute a transnational social class. This idea has been taken up by Sklair (2001), who sees the emergence of a TCC as a class in and for itself. Sklair (2001) identifies four fractions of this global class: globalizing politicians, globalizing bureaucrats, multinational enterprise executives and global consumer elites. Each of these fractions has a vested interest in the health of world-city economies, and the fractions shape urban planning in ways that generate a world-city variety of Molotch's (1976) 'growth machine'.

Local governments' tax revenues are often dependent on the economic health of the local economy to such a degree that politicians may seek to mobilize urban planning to promote real-estate development and the brand of their cities. Bureaucrats deploy legal and financial arrangements, urban planning policies and regulations in order to promote property development. Multinational enterprise executives in all industries lobby on behalf of and seek to gain benefit for their corporations from urban development proposals. Consumer elites have the disposable income to speculate on rising residential property prices. In doing so they have not only fuelled residential construction as part of some of the most spectacular high-rise skylines, but also stimulated distinct transnational residential property and relocation service industries across the Pacific Rim (Olds, 2002) and between world cities twinned by the migration of skilled workers (White and Hurdley, 2003). Elsewhere – as with the outer suburban gated communities and 'transnational houses' of Accra (Grant, 2005) – the effects of consumer elites on urban landscapes may be less obvious and less reported in international media coverage.

The TCC is central in the coming together of local politicians, bureaucrats and 'starchitects' to build iconic buildings and instantly recognizable skylines, in a bid to infuse urban planning with some of the seductive power it may otherwise have lost as a statutory or regulatory activity. Singapore is a prime example of this rediscovered seductive power of the city. Its skyline, along with elements of its planning approach, might be said to have serially seduced city leaders across Asia (Bunnell and Das, 2010). Politicians and other elites are invited to advise on similar projects internationally, creating a circularity or self-referential aspect to the urban planning promoted under the aegis of a TCC.

The activities of a TCC appear intimately connected with the rise of the 'real-estate state': 'a political formation in which real estate has inordinate influence over the shape of our cities, the parameters of our politics, and the lives we lead' (Stein, 2019: 5). The interests of the internationally mobile elites who compose the TCC are interwoven with municipal urban planning, as this is where financial investments and the intense and often uncompromising interests that go with them are made concrete. For sure, 'capitalism never made planning easy – organized money could always thwart the best laid plans – but today's urban planners face an existential crisis: if the city is an investment strategy, are they just wealth managers?' (Stein, 2019: 6). The planning imagination will have to free itself from the corporatized club–state blinkers of this wealth management role if our cities are to be inclusive as well as sustainable.

The effects of the TCC on urban planning are seen in a series of specific planning policies and finance tools that have become significantly diffused. These include the likes of enterprise zones (EZs) and business improvement districts (BIDs). The EZ idea had its origins in the UK in a popular thought-piece (Hall, 1977) and came to be misunderstood, simplified and shorn of important ingredients (Phelps and Tewdwr-Jones, 2014). Nevertheless, the idea travelled across the Atlantic to the US. BIDs – in which 'shadow governments' in the forms of associations of businesses assume partial responsibility for some aspects of the policing, planning and service delivery of defined parts of cities – have travelled in the opposite direction across the Atlantic (Ward, 2006). Tax increment financing (TIF), originating in the US, has been examined and introduced elsewhere as one means for city governments to stimulate urban development. Here, public investments thought necessary to enable private development are funded by speculating on future streams of business rate income as a result of the uplifts in land and property values, as discussed in chapter 4.

Urban planning and the mediation of urban tastes

As Stone (2004: 549) notes, urban 'policy transfer is just as likely to be achieved by mechanisms embedded in markets and networks as in the hierarchies of the state', and this raises the question of the role of a host of market intermediaries in urban planning thought and practice. Indeed, the policy-making process continues to become more roundabout in nature (Majone, 1989), to the point where government has progressively shifted to occupying a meta-governmental role (Levi-Faur, 2005). As we saw in chapter 1, urban planning might be regarded as part of the division of labour in society. To say so may appear to reduce planning to a coordinating activity, but urban planning's intermediary role also signals its seductive potential – a power that may be even more difficult to pronounce on ethically with any confidence now than in the past. The seductive power and reach of urban planning is now less rooted than before in any particular set of urban planning actors – notably state planners – or even the ethical behaviours, practices, decision-making processes and particular tools that have been regarded as the traditional preserve of urban planning.

The majority of professional planning expertise in OECD countries now works in the NGO sector or in service of private sector clients. Thus, the large majority of urban planners themselves have once again become intermediaries in a fundamental way.[2] Here Bourdieu's (1984) ideas of the search for social distinction as they might apply to the production and consumption of the built environment become relevant. The built-environment professions might be considered urban taste makers. In the case of urban planning, this taste-making attribute is being rediscovered as expertise resides outside of statutory planning systems. The role of urban planners as taste makers is implied in the manner in which they bend to consumer desires. It is as well for urban planners to remember these desires, as they play into aspirations to modernization (Ferguson, 2006) and the emergence of particular notably (sub)urban forms (Bruegmann,

2005). Whether such urban settlements are authentic or inauthentic, whether they are sustainable or unsustainable, in some respects is beside the point: it is clear that they reflect important desires, though the seductive nature of some planning ideas poses questions of an apparent prioritization of desires over needs. There might be some concern that 'urban planning is not a key community interest in the transfer business, mainly because urban planning does not have priority among decision makers' (Kunzmann, 2005: 236). However, urban planning thought and practice can seduce by better distinguishing themselves from the selling of urban dreams by architects, developers and real-estate brokers.

In a remarkable turnaround, the urban planning zeitgeist has seen orthodoxies of urban decongestion and suburban planning schemes supplanted by their opposite in new orthodoxies of the compact city and urban density (Neuman, 2005). For these new urban planning orthodoxies, the shocking and unfortunate reality is that suburban forms have only in the past few decades begun to be 'exported' in earnest from the likes of the US. America lacked the history and urbanity of Europe, and its economic and cultural revenge has been the promulgation of a seductive, low-density suburban landscape, its accompanying elements and the urban planning sensibilities and tools associated with it (Beauregard, 2006). While distinctly different and potentially new global suburban forms are visible, likenesses of America's low-density, detached, single-family-home suburbs can be found in the new towns to the west of Jakarta, the multimedia supercorridor of Malaysia, Waterfall City in South Africa and Forest City in Malaysia (Dick and Rimmer, 1998; Moser, 2018; Murray, 2017).

Many of the individual elements of the suburban landscape continue to become near-universal cultural models aspired to variously by individuals, clubs and states. The bungalow is one example. 'It is a dwelling type – possibly the only one – which, both in form and name, can almost certainly be found in every continent of the world' (King, 1984: 2). The regional-scale shopping mall and the modern airport are both creatures of the outer suburbs of major cities and are found not far apart from one another. Though initially failing to live up to their originators' hopes of becoming the new downtowns for suburban America, the format of suburban shopping malls has been copied far and wide. The airport has been actively promoted as an international 'Aerotropolis' model (Kasarda and Lindsay, 2011). Lest we believe the corporate campus (for headquarters and/or research and development) is a thing of the suburban past (Mozingo, 2016), we should note that Apple – that corporate icon of consumer seduction – has just opened a new campus in suburban Cupertino, California.

If the single-family, detached-house, residential suburb had lost some of its shine as an urban form by the 1970s, it is one that is in the process of re-enchantment (Knox, 2008). The enchantment of suburbia involved seduction on a mass scale. Its re-enchantment is based on the differentiation of this form into different market niches. These include the residue of mass markets for lower-middle-income households, an increasing market for low-income or 'multi-unit' apartment housing, and any number of high-value niches such as those catering to 'McMansion' or socially selective New-Urbanism-inspired developments. Thus, New Urbanism, which has gained traction in planning circles in the US, has internationalized to a degree (Passell, 2013). Within the US setting the two New Urbanist developments of Seaside and Kentlands have played important roles in establishing and demonstrating the approach. The planning and development of Seaside effectively launched a 'collective effort to derive and analyse desirable urban qualities and build them into new places, in response to the concurrent depredations of suburban sprawl' (Passell, 2013: 5). The later development of Kentlands in Maryland 'became integral to the ability of the New Urbanism to reproduce itself by serving as proof of its capability' as a model (Passell, 2013: 57). Controversy surrounds the justice and equity credentials of the New Urbanism, since the neotraditional design aspects favoured have actually been a response to some of the premium niches now evident in the suburban home marketplace (Calthorpe, 2005).

Pick-and-mix planning

One implication of the intense exchange of urban planning policy and practice in an age of unprecedented international economic integration, involving significant blurring of the boundaries between states, clubs and citizens, is the emergence of pick-and-mix urban planning.

It is unwise to assume that copying in urban planning practice is inevitably and necessarily problematic. After all, history provides plenty of examples of overt copying that we are grateful for as models of architectural or urban planning virtue and authenticity. There is enough in the history of urban planning for us to be able to point to the significant mixing of styles and principles and a strong measure of hybridity in almost all urban societies. And while Jane Jacobs (1969: 178) was able to describe how 'every city has a direct economic ancestry, a literal economic parentage, in a still older city or cities', it may be futile to search for urban antecedents or origin points (Bunnell, 2015). From our necessarily foreshortened view into the future prospects of urban copies, it is hard to judge the merits of pick-and-mix urban planning. One saving grace of pick-and-mix is that its motives and its fruits will inevitably be contested, adapted and improved.

Portmeirion in north Wales might be seen as a prototype of the sort of pick-and-mix urban planning that may increasingly be upon us. Over a half century or so, architect Clough Williams-Ellis recreated a mix of his most loved architectural styles on a strip of coastal land. Although the conscious architectural pastiche product of a single mind, it carries considerably wider appeal as a tourist destination. Elsewhere, major cities such as Las Vegas – once *the* reference point of the American future of speed – appear instead to have swerved off strip to the pastiche reproduction of iconic buildings. Does 'The Strip' remain a liminal space of detached observation from the car, or is it a place to wander slowly and vicariously immerse oneself in assorted bits of history brought into convenient, walkable juxta-position? In Las Vegas, as in the new capital city of Nur-Sultan (formerly Astana) in Kazakhstan, highly varied architectural styles have been assembled at close quarters.

Elements reproduced from elsewhere can be distributed on a grander scale across a single metropolitan area, or indeed across the metropolitan areas of an entire nation, as in China. Shenzhen itself is home to an eco-tourism resort built to resemble Switzerland's Interlaken. The cores of Shanghai's suburban new towns reproduce the varied architectural styles of European nations, but have from the outset been conceived as part of the marketing effort for entire developments, and themselves command a premium as residential real estate (Wu, 2015). Increased mobility has enabled the gifts of the grand world tour to be brought home or recreated. Today, the experience of global urban variety is enormously facilitated by virtual mobility – a mobility of static and live images of urban scenes and elements provided by the internet. It could be said that the internet provides such a stock and flow of urban images to promote the city ever more as a collage. Wu (2004) provides evidence of how an 'Orange County' model of a gated community developed in the outskirts of Beijing was effectively ripped from its California context and implanted into China without the architects or developers having ever visited the locations and building forms being copied. The marketing of this particular development under the label *linian* – conjoining two words: *lixiang* (ideal) and *gainian* (concept) – seems only to highlight how detached in time and place the urban planning process itself was in this instance.

Citizens: planning for ordinary needs

The coercion and universalism associated with states and corporations as clubs discussed up to this point are both perhaps most closely connected with a class of world cities from which political and economic power continues to be projected. However, the large body of academic research describing the system of world cities and some of their distinctive urban planning challenges has come under criticism. 'The global/world cities framework asserts a hierarchy of cities but is unable to account fully for the materialization of such a hierarchy, and even less so in relation to the long histories of colonialism and imperialism. Space is a "container" ... its "production" remains unexplained' (Roy, 2009b: 824). The world-cities literature fails to offer much insight into the 'ordinariness' (Robinson, 2006) and 'worldliness' (Roy and Ong, 2009) of most cities as sources of urban planning invention and adaptation in the erstwhile urban and economic periphery of the world economy (Nasr and Volait, 2003: xi).[3]

Moreover, something of a clash of urban planning rationalities is apparent in cities across the global south. Here statutory 'planners ... are located within a fundamental tension – a conflict of rationalities – between the logic of governing and the logic of survival' (Watson, 2009: 2268). The tension may not be exclusive to the experiences of statutory planners working in cities of the global south, since, as we saw in chapter 3, informality appears in the global north. Nevertheless, a large part of the urban population of global south cities lives in what the UN defines as slums. It is unclear how the ordinary needs of and challenges facing the many informal households in cities of the global south are served by the concern of modernist, global north planning with 'the aesthetics (of order, formality, geometric simplicity), efficiency (separation of land uses, and the free flow of traffic) and modernisation (slum removal, high-rise buildings, formal green spaces). As a result, 'a significant gap has opened up between increasingly technomanagerial and marketized systems of government administration, service provision and planning (including, frequently, older forms of planning) and the every-day lives of a marginalised and impoverished urban population surviving largely under conditions of informality' (Watson, 2009: 2260).

This tension between competing rationalities that urban planners find themselves in is nevertheless one of justice, which urban planning thought and practice have typically not shied away from. The problems of seeking to exceed modes of survival are rarely amenable to simple, universal solutions, as illustrated in the failure of Hernando De Soto's ideas regarding the mobilization of assets of the urban poor through the formalization of land and property titles. Indeed, urban informality often presents opportunities for big business as well as ordinary citizens to benefit from land-value uplifts and development arrangements, in ways that mean that outcomes at this 'frontier of accumulation' (Roy, 2009b: 826) defy easy generalization.

What is lost in global north understandings of urban planning and its potentials is captured in the thought that

> theories of urban modernity ... have drawn a stark line between 'modern' cities and other kinds of cities, variously described as Third World, perhaps African, perhaps developing/underdeveloped, perhaps colonial. At best these 'other' cities have been thought to borrow their modernity from wealthier contexts, presenting pure imitations rather than offering sites for inventiveness and innovation. (Robinson, 2006: x–xi)

The problem here is that the cosmopolitanism that is seen to drive invention and innovation may be too readily and exclusively associated with world cities of the global north. For 'if we ask ... where the first historical occurrence was of what today we call the "modern multi-cultural" city, the answer is certainly not in the European or North American "core" cities of London, Los Angeles or New York, but probably in one-time "peripheral" ones of Rio, Calcutta, Shanghai or Batavia' (King, 2004: 74). Instead, then, Robinson (2006: 7) posits 'a concept of the modern that sees many different cultures in many different places as enchanted by the production and circulation of novelty, innovation and new fashions'. Cities of the global south are ones in which informality, far from being a problem to be planned away, can be a source of new urban planning thought and practice. Here, then, the challenge of leveraging the potential while ameliorating the problems of global south cities is one on which hopes for new repertoires of urban planning regulations, policies and practices rest.

Participatory budgeting (PB) is one innovation from the global south that has been celebrated, widely reported upon and diffused as a policy-making method, in the sphere of urban planning in particular. It originated from Porto Alegre in Brazil in the 1990s as a pragmatic means of involving citizens in the difficult task of allocating scarce public resources. PB involves citizens in deciding priorities for the municipal budget and is credited with opening up municipal governance to new forms of public participation, and even with helping to fashion a new kind of redistributive, bottom-up democracy. This model urban planning practice has travelled far and wide. The PB story is one that might be thought to reveal the individuals, institutions and structures that Sutcliffe (1981) saw mingling in the processes by which urban planning thought and practice have circulated historically. For example, its arrival in Maputo, Mozambique, can be read in terms of the efforts of advocacy groups, networks and particular individuals, and as an initiative driven top-down by international organizations like the World Bank (Carolini, 2015). It has been the inspiration behind NGO Yayasan Kota Kita ('Our City Foundation') in its work within the Indonesia city of Solo, where the participatory mapping of local resources and assets has been a vital input into empowering citizens to participate in a more informed manner about local government policies and priorities. As Peck and Theodore (2015: 182) note: 'For a while, in the early 2000s, "learning from Porto Alegre" was the thing to do, particularly, it seemed for left-leaning local and regional government in Europe'; it was a place of 'policy origination' regarding PB. Estimates suggest that PB has materialized as a method in somewhere between 800 and 1,500 cities worldwide (Carolini, 2015), though in the process it has been adapted and deployed in ways that have been seen to dilute or undermine its progressive intent.

A second and obvious facet of innovation emanating from citizen–state interactions and circulating among cities of the global south is related to the older notion of the 'appropriateness' of technology applied in the name of development (which would include urban planning 'solutions'). The rapid bus transit system of Curitiba, Brazil, garnered much publicity as a cheaper and more appropriate solution to the physical and fiscal conditions of global south cities than expensive overground or underground tram, light rail and metro systems. The issue of the appropriateness of urban planning expertise is apparent in the manner in which African nations have begun looking more to South Korea than, for example, Singapore for new town planning and construction (Kim et al., 2021).

The idea of appropriateness of technology finds a partial echo today in ideas of frugal innovation (Knorringa et al., 2016). This sometimes emphasizes the mobilization of the club resources of multinational enterprises in the generation and diffusion of low-cost solutions to a plethora of problems relevant to urban planning problems across the global south, though it remains ideologically contested. Certainly, citizens are every bit as able to match the cost-containment capabilities often claimed for clubs and states (Anzorena et al., 1998). Indeed, multinational enterprises, as club actors, have been acutely aware of the cost-containment and revenue-generating possibilities associated with frugal/base-of-pyramid innovation, to the point where the social benefits of the notion can come into question (Meagher, 2018).

Conclusion

The geohistorical sensibilities of the urban planning imagination alert us to the fact that there has been a mix of actors – citizens, clubs and states – and a mix of forces driving the exchange of ideas internationally. Both military and economic coercion of states have been prominent in the past of imperialism. Imperial expansion ushered in rational-voluntaristic forms of exchange of urban planning ideas that are a major part of the professionalized and bureaucratized apparatus of urban planning now partially decoupled from nation states. The products of this bureaucratization of urban planning are multiple, and exceed well-rehearsed national repertoires of statutory urban planning to encompass mixes of not only state and club but also state and citizen imaginaries.

The seductive power of the urban planning imagination has been apparent since ancient times and continues to be in evidence. It is a power that may yet confer distinction on urban planning in the thoughtful, practical and tasteful making of good or better urban places. Some of this seductive power is to be seen in instances of unthinking imitation or juxtaposing of numerous diverse elements, which may be a cause for concern. Cities of the global south are home to innovative and productive citizen–state and citizen–state–club mixes of the urban planning imagination that may be important to its future more broadly. However, within an increasingly complex division of labour, every ounce of urban planning's imagination – its wisdom and its power to seduce – will be needed if it is to be more than just a prop to the 'real-estate state' or exert anything other than a marginal effect on the future shape of our cities.

8 Prospects: what is the future of planning?

Introduction

That 'planners seem to share a common style of thought, irrespective of where they are or what they plan' might be thought to hold in an era of globalization and the emergence of elements of global governance (Reade, 1983: 159). Twenty or so years later, such a common planning culture appeared lost to Sanyal (2005b) in a world of diverse planning systems and cultures, and of an increasing mix of urban planners as citizens, clubs and states deploying a diverse range of methods. The accusation that urban planning has ignored knowledge in other disciplines might be nearer the mark, in the thought that 'planners must not expect lawyers and economists to learn their language. ... planners' language is often less precise than that of lawyers and economists. So it is the planners who must learn to use a new language' (Needham, 2006: 6).

Part of my concern in this book has been to contextualize an understanding of what planning is, what it does, how it does it and with what success. Planning's language may emerge as less precise than that used in these other disciplines, and yet its value – its necessity – emerges all the more clearly when placing it in its wider context. That is, urban 'planning is naturally positioned at the interface of a variety of disciplines in a brokering position, connecting and linking disciplines through its unique lenses of spatiality and action' (Frank and Silver, 2018: 245). 'We plan because our collective experience suggests that things work out better than if we leave them alone' (Hanson, 2017: 262), and that collective experience is found in insights condensed from across the disciplines. I have highlighted the close relationships that have existed and should continue to exist between historical and geographical thought and planning thought and practice. Urban planning is central to our sense of being in the world. It is 'a way of knowing the world as well as a way of remaking it' (Stein, 2019: 7).

The fact that many urban planning curricula contain a significant element of its history not only identifies the importance of an historical perspective in connection with the substantive concerns of urban planning (chapter 3), but also suggests the value of historical perspective as part of the methods of urban planning (chapter 5) (Abbott and Adler, 1989; Ward et al., 2011). Early planning histories enabled planners

> to explore issues such as the density and massing of buildings, street alignments, perspective and the ordering of public space. Historical awareness also allowed planners to understand the evolutionary processes of urban development, the functions of urban places and the extent to which natural and human forces interacted to shape and change the visual and functional character of the city. (Ward et al., 2011: 233)

'The new planning history' promises both a more analytical perspective on evolution – continuity and change – as discussed in chapter 2, and a more critical engagement with the sorts of structures and mechanisms driving the likes of international exchange of urban planning ideas, concepts and methods that I discussed in chapter 7. The history of planning methods is important too, since 'knowing more about how earlier plans were prepared, and what aspects worked well, might assist planners in choosing a methodology for new plans' (Batey, 2018: 47). Equally, of course, there is the real danger – given the churn apparent in urban planning policy – that valuable plan-making experience and wisdom might be lost.

A better understanding of place is still something needed in planning thought and practice (OECD, 2017b). Here I have emphasized the need to recognize the value of the urban planning imagination in fostering a relational or 'global sense of place'. This is necessary in an urban planning world that continues to increase in complexity.

> Spatial and land-use planning today is intimately connected to much broader agendas such as the transition to a low carbon economy, reducing social spatial inequalities, and creating opportunities for economic prosperity. Spatial planning is therefore linked to policy ambitions at multiple scales, extending across sectoral issues and involving an ever wider array of actors in structures of governance. (OECD, 2017b: 68)

In this book I have reduced some of this complexity to the activities of three sets of planning actors, and it is worth considering the strengths and weaknesses, possibilities for change and value of striving for distinctive and place-affirming or place-improving mixes of the imaginations of those actors.

The limits of citizens, clubs and states as urban planners

It is unwise to attribute essential rationalities to citizen, club or state as urban planning actors and there are limits to the imagination, wisdom, resources and methods that can be mobilized by each of the three groups of actors – citizens, clubs and states – in isolation.

Citizens

Is it time to more fully license and empower citizens? Citizen urban planning exists in the minority pursuit of dream homes, teardowns (demolition and replacement with typically larger homes on generous lots) and so forth that are taking place in some of the most highly formalized global north planning systems, but should it be encouraged as a mass pursuit by states, even if within limits? Thought experiments (Banham et al., 1969) and instances of non-zoning such as Houston in the US (Siegan, 1970) suggest that the counterfactual of less or non-planning planning may produce results little different from planning in physical appearances. The charge here is that the increasingly complex bureaucratic nature of planning as a primarily statutory activity has not only failed to live up to many of its place-making ideals, but also stifled the energies of citizens as they participate in the production of the built environment. Non-plan ideas are sometimes misunderstood but can be read as limited, experimental desires to let loose the better aspirations of citizens. Those possibilities remain alive and well today, as the resourcefulness of communities in cities across the global south testifies. However, in the global north we have yet to find the planning courage or the appropriate tools to facilitate this in ways that can reinvigorate some of those cities that continue to struggle economically, on the one hand, or produce better, new suburban places than much of those produced by or for clubs, on the other hand.

Should plan making, planning decisions and bargaining with developers be devolved to local citizens? Would citizens strike better deals than state planners or their elected representatives do? Citizens' intimate knowledge of and intense interest in the fortunes of their neighbourhoods might help fashion better places from the user perspective than the plans conceived by state or club actors for them. That same local lay knowledge allied to self-interest might make for bargains that leverage better economic, social and environmental outcomes from developers than can typically be achieved by disinterested elected representatives and state and club planners. Statutory planners' role might change to that of expert, experienced activist advisors to or advocates for citizens, rather than that of disinterested intermediaries or organizers of communicative or deliberative processes. Instead of being neutral intermediaries, professional urban planners might occupy a position firmly on the side of citizens in negotiations with developers. There is a lot for the state planner to do here in some global north settings. Having taken this sort of responsibility out of the hands of citizens, it is quite apparent in the UK's experiment with localism to date that it will take time for citizens to reacquire the perspective and skills that are needed.

Across the global south, the do-it-yourself incrementalism of citizens' urban planning practices, in the form of squatting, repair, consolidation (Bhan, 2019), and innovative solutions to budgetary priorities at the intersection of citizen and state actors' capabilities, already appear to be of wider relevance to and resonate partially with yet other disparate practices of protest found in cities in the global north (Iveson, 2013). After all, some of the planning systems and cultures reviewed in chapter 6 have a distinct orientation to the incremental in the urban planning imagination and the actual building of cities. Citizens across the global north continue to take matters into their own hands, proving the relevance of global south urban planning repertoires such as squatting (Chatterton, 2002; Vasudevan, 2015), in protest at the portions of housing stocks that are un- or under-occupied or at the exclusive urban regeneration and gentrification agendas of clubs and states.

Planning in the name of clubs

Much of the excitement and monetary and experiential rewards open to professional urban planners lies outside statutory planning, in the private or NGO sectors, in consultancies or lobby or social support groups that serve often particular individual or club 'client' interests.

This is a fast-moving field in which group work and visual creativity, in particular, come into play. In the best case, the NGO-sector urban planner may play a vital role in ensuring that planning alternatives are properly aired in the face of TINA ('there is no alternative') claims, and in ensuring that the weakest in society are represented in planning deliberations and decisions on development proposals.

Planning in the name of clubs is often – though in rather utilitarian ways – strong on a certain sense of public participation. In the sense that developers have a vital interest in ensuring that their schemes are what buyers want, they are sensitive to consumer demand at the outset. Some take care to respond to user or occupant experiences and provide some form of customer 'aftercare'. Club planners, it would seem, may be more comfortable with participation and citizens' desires than statutory planners – though this itself can represent a tyranny of passive response to and amplification of market demands, to be avoided in urban design (Carmona, 2011).

Where those clubs producing, managing or occupying urban space are relatively broad – perhaps in the form of privately developed new towns – there may be value in turning over planning powers (such as eminent domain in the US or compulsory purchase), within limits and under clear rules, to private sector developers and occupying citizens. The aim would be to deliver large, socially mixed and partly self-contained or self-sufficient communities that the state sector is often now unwilling to countenance because of the political fallout, or unable to consider due to lack of finance.

The urban planning imagination is especially important in resolving problems of urban development under one or more of the conditions of uncertainty, irreversibility, interdependence and indivisibilities (Hopkins, 2001). These collective action problems are precisely those that draw upon all the experience and knowledge gained by planners across their careers. However, while state planners perhaps ought to be licensed to act as advocates (Davidoff, 1965), their entanglement in an urban planning division of labour means that it is often they who are blamed for urban blight. They rarely have the capacity for what are lengthy and complex negotiations, are not really licensed to engage in such negotiations, and might not be trusted to take part, given some of the spoils involved. There may be benefits, then, in licensing particular club – consultant or NGO – urban planners to solve numerous small-to-medium-scale collective action problems that result in inertia in planning and development processes. Examples are long-neglected or 'blighted' sites, complex private hold-outs on otherwise desirable development, and the persistence of 'zombie subdivisions'.

Yet planning of, or for, clubs is urban planning in the service of often particular client interests. In the worst case, the private sector consultant planner can emerge as something akin to the most cynical of lawyers, interested in winning an argument at all costs, and for hire to the highest or any bidder regardless of their place-making intentions. Pervasive planning for our cities in the form of clubs risks balkanizing them into many fragments or enclaves, undermining all that we most appreciate of great cities as a meeting of means and minds. For the majority of the time, planners acting on behalf of NGO clubs are likely to exhibit neither the worst nor the best of human nature. Still, given the rise to prominence of many NGOs within world society, NGO planners can themselves become de-radicalized as part of the 'business as usual' of bureaucratized and formulaic urban planning, including the universalization of particular development or SDG 'models'. Ultimately, while acting for NGO clubs, professional urban planners may do little to challenge urban planning orthodoxies that have outlived their useful life or are indiscriminately applied. The challenge remains for planning processes to be designed in ways that let the best nature and capacities of club planning come to the fore.

One peculiar club of sorts – but one whose presence could have been more to the fore in my discussions of the urban planning imagination – is academia. Is it time to reconsider the role of universities? As they continue to educate urban planners destined for the state sector, syllabuses will probably need to ensure that skills of leadership and social entrepreneurship become more developed than they presently are. Is there a place for urban planning degrees to reinvent themselves with something of the cachet of an MBA/MPA (Master of Business/Public Administration) or a PPE (Philosophy, Politics and Economics) degree? Few city leaders – mayors and chief executive officers – are trained planners, but more than ever they need a broad and synoptic sense of the possibilities open to them in the shaping of the futures of cities, as well as access to the historical wisdom offered in urban planning. In this respect, the expanding scope of urban planning might be less a cause for professional angst than a source of influence, since the breadth of planning expertise is certainly no wider than that of elected political leaders (Davidoff, 1965).

A re-evaluation of the civic mission of universities has been under way for some time now (Addie, 2017), but little of this sees the university as anything other than a producer and dispenser of unrivalled knowledge, rather than an institution with faculty licensed and encouraged to animate some of the interactions at the centre of figure 2.1 (see chapter 2). Universities and academics have played these important roles before, as I discussed in chapter 5, and they can be relicensed to do so in the future in universities that are less like the for-profit corporations they have become, and as they no longer educate and train urban planners who are destined primarily for the state sector. Some sense of the possibilities has been showcased in the Newcastle City Futures initiative, where universities and university scholars were instrumental in the use of exhibitions to generate public interest in planning (Tewdwr-Jones et al., 2019).

W(h)ither the state?

As revealed with respect to planning across the global south (Parnell et al., 2009), it is the case more generally that, while the value of planning is reaffirmed, what planners ought to be doing and where the urban planning imagination is located are in need of significant reappraisal. Much of what we associate with urban planning is that which takes place by law in the state sector on behalf of national and local governments in order to regulate the spatial distribution of activities, use of land, and type and standards of buildings. This statutory planning remains important for the often residualized sense of the public interest that is associated with it. However, it has long been apparent how slow plan making is and how slow to take effect plans may be – if indeed they have the effects intended.

It may be less appreciated just how constrained statutory planning has become. The pervasiveness of statutory urban planning gives it the appearance of a bureaucratic tyranny (Carmona, 2011) to be overcome in the practice of designing places. However, the real power of statutory planning rarely exceeds that of saying no to development (McGlynn, 1993). Its responsibilities and ethics of care overburden both the system and individual planners. Where urban planners in the state sector were able to exercise discretion in some of the planning systems and cultures reviewed in chapter 6, they no longer appear able to provide the imagination, the vision or the oversight of cities whose development may increasingly be turned over to the twin codes of zoning (Talen, 2011) and computer software (Kitchin and Dodge, 2011). Instead, 'groups of individuals other than planners determine the future of cities with planners now acting largely as brokers or negotiators' (Batty and Marshall, 2012: 37). Statutory planners are now largely unable to really deploy their experience, creativity or vision, and this is surely not desirable. Worse, the accusation can be levelled that statutory planning has in many instances been shorn of any of its transformative credentials. 'Good' planning increasingly seems to be indistinguishable from real-estate development (Molotch, 1976) and generalized gentrification (Stein, 2019), as we saw in chapters 4 and 7.

Has statutory planning become drained of imagination? In this regard, 'the extent to which planners operate within their statutory role or extend beyond it is in large part a function of political culture, social trust and capacity' (OECD, 2017b: 43). Yet 'the freer planners are from formal constraints, the more economical and inventive they are *able* to be, and the more they are *able* to invite participation in their work' (Stretton, 1975: 313). The rehabilitation of the reputation of the statutory planner is a matter of importance for all of us. Letting go of political business as usual and seeing less like a state will be important things for governments to do.

Should state sector planning be allowed to wither? If the urban planning profession is thought insecure, it is worth reflecting that it is older and no more insecure than that of management consultants (McKenna, 2010) or investment arbitrators (Dezalay and Garth, 1996), who have managed to carve out prominent places for their expertise. Professional urban planners have emerged outside of the state sector as the arbiters of urban taste (chapter 7). For sure, the urban planning profession does have 'a provisional and fluid identity precisely because it embraces dynamic currents of belief' (Hoch, 2019: 45), and that will mean that it continues to span the state–club intersection and it may yet become developed at the state–citizen and club–citizen intersections.

In many global north contexts, it is clear that 'marvelous planners were not trained as professional planners and many planners with professional training have had little influence on their surroundings. The difference between the two is the sometimes mysterious quality of leadership' (Garvin, 2009: 25). Yet the 'ordinary' local government planners in liberal market states need to be licensed to exercise a greater measure of discretion and responsibility, especially those seasoned public servants whose often vast experience and presumably unheard urban planning innovations need to be aired and tested. The synoptic planning imagination remains vitally important, since the 'art of enhancing the life of the city' is something more than merely architecture, design or surveying (Geddes, 1904). Notably, 'Where planners strive to connect the dots to show interrelationships among interests, politicians prefer that they be compartmentalized to avoid inconvenient coalitions that upset established patterns of power' (Hanson, 2017: 11). As skilled generalists, urban planners are integrators of ideas (Garvin, 2009). The entrepreneurial role is not one that planners originally envisaged for themselves (Levy, 2016: 100), but a measure of innovation and even bureaucratic entrepreneurialism associated with urban planning can usefully be nurtured.

The city as laboratory: productive mixes of the urban planning imagination?

If urban planning is cast as a 'practical minded activity that attends to purpose, context and use more than principles of rationality and method' (Hoch 2019: 1), then a mix of actors' motivations and expertise in the urban planning imagination in the shaping of places may not seem such a bad thing. In one version, it might promise the best attributes of citizen, club and state planning actors combined. However, today's reality appears more likely to involve quite limited combinations of the attributes of these different actors and not necessarily their best attributes. There is a distinct danger that pick-and-mix urban planning will lead to the proliferation of bad or poorer places – a case of cities emerging as somehow less than the sum of citizen-, club- and state-actor parts. There is the danger that the unthinking business-as-usual planning of each of these actors becomes combined in a lowest-common-denominator urban planning deriving from an eclectic mix of imaginations, methods, resources and substantive concerns. Whatever the future of urban planning, I would certainly want to advocate avoiding such poor instances of pick-and-mix.

Contemporary urban planning – especially that concerned with promoting the social, environmental and economic sustainability and resilience of cities – increasingly appears to involve elements of experimentation (Bulkely and Castan Broto, 2013), which in turn appear to involve new mixes of citizen, club and state actors.

Indeed, the value of a plurality of experiments is one lesson drawn from examining some of the UK's most successful plans (Wray, 2016). These elements of experimentation include the conception of entirely new settlements as test beds, the piloting of policies in neighbourhoods, the search for best practices, and networks of comparative learning between places (Barnett and Parnell, 2018: 32). If urban planning is an act involving 'all those who need to learn about their environments – public or private, social or natural – in order to change them' (Forester 2006: 124), these experiments signal desires and opportunities for learning. Limited experiments might be one means of testing out how productive mixes of the imagination, methods, resources and substantive concerns of different planning actors might be achieved and generalized to some local circumstances. They might also simply exist as so many projects which exist in a world parallel to and with little bearing on, for example, statutory planning. Or they might see the city made over by some of the worst natures of our three sets of planning actors.

Organicism – the likening of the spatial configuration and functioning of the city to that of an organism or organisms – has recently been supplanted to an extent by the metaphor of the city as a laboratory or site of experiment. It is surprising that this metaphor has not drawn much academic criticism, given the question of who is experimenting on whom. Lawhon et al. (2020) note the difficulties of producing urban planning knowledge and engaged theory in contexts where citizens recognize all too well that they are the subjects of research on urban planning interventions. Academic and professional urban planning practitioner views – many of which are highly cosmopolitan in orientation – will need to guard against the urban planning imagination being tied to so many shallow, short-lived endeavours out of which valuable knowledge is gained by clubs or states but where citizens may gain little or nothing. There is a real danger that the city as a laboratory becomes no more than a series of interesting but time-limited, disconnected and non-cumulative projects.

There is a more particular danger still of the city becoming a controlled laboratory for a fusion of state and club experimentation. Small nations or city states, such as Singapore and Dubai, which manifest an apparent mastery of total formal physical and social urban planning, have risen to prominence in the global urban consciousness. They have done so 'at home' with a measure of the sorts of social control exercised by former imperial powers in their colonies. This is troubling to the extent that they present a seductive model to politicians, club leaders and some citizens of the potential control of otherwise messy and wicked urban problem issues – not least the respective roles and freedoms of each of state, club and citizens as participants in the making of cities. Do these cities represent a combination of citizens, clubs and states that is somehow less than the sum of its parts? To the extent that these city states are a strange and near-complete fusion of very particular state and club imaginations, we must hope that they represent minority secessions from a world of genuinely worldly cities.

However, on the whole, the laboratory metaphor has been used in positive ways that embrace the diversity of populations and physical settlement forms in proactive attempts to shape the future sustainability of our cities. We need these experiments, as they continue to offer potentially scalable alternatives in the face of statutory urban planning that all too commonly now presents us with TINA development proposals and plans. 'We must ... assume that our cities are malleable, and that we – citizens, administrators, architects, planners – can do *something* to make our preferences clear, and that we have only ourselves to blame if things get worse rather than better' (Rykwert, 2000: 7). To the extent that state planners are able to operate outside of statutory constraints, to the extent that citizens can be brought into state planning exercises, and to the extent that the innovation and resources of clubs can be put to better or best use in the service of broader publics, pick-and-mix planning may be greater than the sum of its parts. I have sprinkled reference to some examples of these overlaps in the imagination of actors – the overlaps in figure 2.1 in chapter 2 – throughout this book as evidence of the possibilities that exist for urban planning in the future. They are by no means exhaustive of the experimentation that has existed and will continue to take place.

Conclusion: urban planning's disciplinary dialogues

In championing urban planning, I have been championing a closer relationship between the planning history and geography disciplines than has been the case in recent decades (Phelps and Tewdwr-Jones, 2008). My privileging of history and geography is not meant to exclude dialogue between planning and other disciplines such as architecture, landscape and urban design, sociology, and property and real estate economics. As the stock of planning knowledge, wisdom and methods continues to expand, the urban planning imagination will probably need to draw on other cognate disciplines, such as development studies and anthropology, where important place-based analytical insights and practical applications may be found, and where some of the language needed to rethink urban planning from and for the global south is being developed.

The geohistory of exchange indicates how the distinct urban planning systems and cultures of nations described in the previous chapter are constantly acted upon and altered. As such, the urban planning imagination is one that, of necessity, has variations on common themes. If a universal and historically stable core to something called urban planning cannot be defined, Patsy Healey (2012: 200) nevertheless mused on the relativity of that core set of interests: 'The idea [of planning] instead carries with it a normative orientation ... honed by a continual interaction between situated practice experiences and theoretical development. It is contingent both in the way ... debates develop and in working out how the idea may inspire particular practices to develop.' In this, urban planning 'cannot ignore the [re]construction of the institutional systems on which the implementation of its policy propositions will depend' (Parnell et al., 2009: 238). Likewise, 'theory can and should travel outside its original context to be shaped, challenged and reformu-lated across space' (Ernstson et al., 2014: 1564), and this is another effect of the international exchange in which the urban planning imagination is entangled.

An increasingly important task for urban planning practitioners and educators will be to consider the possibilities for and meaning of emerging geographies of the urban planning imagination. Will an Atlantic era of urban planning foment and exchange dating to the late 1800s (Rodgers, 1998) be eclipsed in the present century by a Pacific era or a Sinicization of urban planning, and to what effect? A Pacific arena of exchange, partially underpinned by the trade links promoted under the Comprehensive and Progressive Agreement for Trans-Pacific Partnership, could draw sustenance from, or may founder on, the greater diversity in urban conditions, substantive concerns and institutional contexts than that encountered in transatlantic exchanges. Any Sinicization of urban planning may do little for the cause of the 'slow planning' needed to combat the policy churn and loss of institutional memory, or urban planning already reduced to servicing real-estate interests, now apparent in many a global north nation. Can south–south exchange of urban planning thought and practice (Harrison, 2015) bridge the cracks in south–south economic and international relations that have emerged (Gray and Gills, 2016)? Will the numeric importance and sheer necessity of citizen-centred rationalities of informal urban planning dictate that elements of 'southern critique' and reconstruction of practice increasingly coexist with, or talk past, established global north planning thought and practice without altering them? Will yet more heterogeneous global south-to-north patterns and processes of urban planning exchange emerge as a result of shared histories, migration, or even some of the partial parallels in the national economic development or welfare-model institutional settings I noted in chapter 6?

In any such discussions, the urban planning discipline is important to the likes of geography, in particular, when forcing a stronger measure of relevance. The question of policy for whom (Harvey, 1974) should not absolve us from the responsibility as geographer-planners to suggest practical courses of action. The Marxist hoping for societal revolution may be about as likely to achieve substantive transformations of people's city lives as the communicative planner endlessly deferring action in search of the ideal speech situation. A greater portion of geographical theory could usefully have traction on substantive issues facing those who plan – citizens, clubs and states – and seek to produce more persuasive rhetoric and visuals with which to influence decision makers. Academic geographers often retreat into a role as analysts, reluctant to offer possible planning sugges-tions, perhaps for fear of attracting the sort of criticism reserved for practising planners. This is a shame, since it is likely that there is continuing value in the practical implications of theory to be drawn out in appropriate venues, such as in the CDP initiatives in the UK of the 1970s (chapter 5), and in the staple of academic book and journal publications.

My concern to bring geography, history and planning together is also one of recognizing the complementarities between the different comparative insights and methods of analysis; together these discip-lines reveal continuity in change and variations on common geographical themes in urban planning thought and practice. In particular, a sense of humankind's fundamental connectedness in thought and action across time and geographical contexts is revealed in the way we make our cities. In practical terms, the complementarities that exist between planning, geography and history, I believe, demand 'a sensitivity to the disjuncture between the assumptions policymakers are forced to make about aggregate conditions and the lived realities they are simplifying' (Pieterse, 2008: 128–9). Modesty regarding the completeness and generalizability of urban planning interventions is something that is quite compatible with striving to make better cities as the embodiments of our being in a world with others.

Notes

Chapter 1 Introduction: what is planning?

1 Much of the critique of modern urban planning as everything and nothing (Wildavsky, 1973), difficult to define (Reade, 1983) or no better than non-planning (Banham et al., 1969) is based on a very narrow view of planning as statutory planning.

Chapter 2 Imagination: what is planning's spirit and purpose?

1 Indeed, the problem for our discussion of urban planning in the present age is that it must take place in a context in which at least one prominent historian believes that no well-established and valuable general propositions currently exist at the world-historical scale (Tilly, 1984).

2 Business cycles reflect fluctuations in aggregate demand over the medium term while long-wave or Kondratieff cycles reflect long-term movements in commodity prices over a period of fifty years or so.

3 BREEAM (Building Research Establishment Environmental Assessment Method) and LEED (Leadership in Energy and Environmental Design) are systems created in 1990s in the UK and the US, initially to certify the sustainability of buildings.

4 https://merics.org/en/analysis/mapping-belt-and-road-initiative-where-we-stand.

Chapter 3 Substance: what are the objects of planning?

1 The materiality of informal settlements improves over time to exceed slum conditions. Rio's favelas 'are built of concrete, reinforced steel, with indoor plumbing, water, and electricity in over 90% of homes', yet their citizens are held in limbo with regard to legal title to land and access to services and opportunities (https://consortiumforsustainableurbanization.org/2020/01/rio-de-janeiros-sustainable-favela-network-and-its-take-on-the-sdgs).

2 https://www.citylab.com/environment/2019/04/cape-town-water-conservation-south-africa-drought/587011.

3 https://www.cbsnews.com/pictures/the-most-polluted-cities-in-the-world-ranked/1.

4 https://www.theguardian.com/environment/2015/jul/15/nearly-9500-people-die-each-year-in-london-because-of-air-pollution-study.

5 The systemic pervasiveness of automobility is further defined by Sheller and Urry (2000) in terms of six features: (1) the quintessential manufactured object produced by the leading industrial sectors; (2) the major item of individual consumption after housing; (3) an extraordinarily powerful machinic complex; (4) the predominant global form of 'quasi-private' mobility that subordinates other 'public' mobilities; (5) the dominant culture that sustains major discourses of what constitutes the good life; and (6) the single most important cause of environmental resource use.

6 Center for Neighbourhood Technology's Housing + Transport Affordability Index: https://www.cnt.org/tools/housing-and-transportation-affordability-index.

7 https://www.ellenmacarthurfoundation.org. The RESOLVE acronym stands for: regenerate the health of the ecosystem through the likes of shifts to renewable resources; share resources and products among different users; optimize by reducing waste in production and supply chains; loop – close loops in the reuse, recycling, recovering and remanufacturing of products, materials and components; virtualize by dematerializing the provision of products and services; and exchange – replace products and services with those that are less resource intensive.

Chapter 4 Wisdom: what does planning teach us?

1 Dick Whittington (c.1354–1423), Lord Mayor of London, and the subject of a folklore tale of how he escaped poverty to become a wealthy merchant.

2 See, for example, https://www.theguardian.com/commentisfree/2020/jan/11/how-first-australians-ancient-knowledge-can-help-us-survive-the-bushfires-of-the-future.

3 See, for example, the 'global suburbanisms' major collaborative research initiative located at York University, Toronto: https://suburbs.info.yorku.ca.

Chapter 5 Methods: what are the means of planning?

1 The McKinsey summary. https://www.mckinsey.com/business-functions/strategy-and-corporate-finance/our-insights/the-use-and-abuse-of-scenarios. In the details of the method of developing scenarios, there

are dangers associated both with discarding scenarios too hastily and in particular dispensing with tails or outliers of a distribution, and with including scenarios where the uncertainty is so great that they cannot be built reliably.

2 https://www.mckinsey.com/business-functions/strategy-and-corporate-finance/our-insights/the-use-and-abuse-of-scenarios.

3 https://consortiumforsustainableurbanization.org/2020/01/rio-de-janeiros-sustainable-favela-network-and-its-take-on-the-sdgs.

Chapter 6 Comparisons: what are the global variations in planning?

1 Booth (2011) also itemizes: the development of theory, the explanation and interpretation of social phenomena and the description of social reality.

2 These are: (1) the primary purposes of local government; (2) local government importance; (3) local government institutional structure, form and setting; (4) decentralization and fiscal federalism; (5) local autonomy; (6) local democracy; (7) local government capacity and performance; (8) local government service delivery; and (9) local government politics and policy.

3 Table 6.1 contravenes one key rule for formal classifications in that it is not mutually exclusive and exhaustive of cases (Wolman, 2008).

4 The term 'Brexit' refers to the referendum in favour of Britain leaving the EU which was eventually enacted and took effect from late 2019.

5 Much of the remaining discretion in the English system within the UK may be reduced if proposals for adopting a new, more rules-based, system akin to that in the US and mainland European nations are adopted. Proposals note the uniqueness of the discretionary system but also emphasize its effects in generating complexity, uncertainty and loss of trust in the planning system (Ministry of Housing, Communities and Local Government, 2020: 11–12). Critics note not only the contradictions in the proposals, such as the short time horizon (of fifteen years) relative to the likely length of time needed to prepare binding rules-based local plans (Bäing and Webb, 2020) but also the claims regarding the efficiency and effects of the discretionary system (Booth et al., 2020).

6 https://aiatsis.gov.au/explore/articles/aiatsis-map-indigenous-australia.

7 Comparative analysis highlights the socially constructed nature of housing found in different countries and how this shapes supply and prices of different housing types and tenures, i.e. the substantive urban planning concern of 'shelter' that I covered in chapter 3.

8 The comparative literature points up important contrasts in the concentration and organization of businesses, including the likes of industrial

relations, and is notable for emphasizing that distinctly transnational business systems have not emerged even in an era of globalization. Different types of business activity – by sector, function or corporate culture – are produced from but also are attracted to qualitatively different business systems, including differences produced by property markets and planning systems.

9   Home (2015) identifies five elements present in urban planning within Anglophone Africa: (1) rules derived from the Caribbean slave colonies centuries earlier; (2) Benthamite utilitarianism applied within colonial administrative institutions; (3) military and hygiene imperatives of cantonment regulations adapted from those developed in India; (4) concepts of dual mandate, trusteeship and indirect rule; and (5) a legal approach to land tenure based on private property and first instituted in Australia.

Chapter 7 Exchanges: what are the global connections in planning?

1   My reference to the plan as theft relates to French anarchist Proudhon's thought that 'property is theft'.

2   A number of different terms have been used to speak to the rise of intermediary actors in the production and consumption of the built environment. Olds (2002) refers to the importance of a 'global intelligence corps' in the promotion of urban mega-projects. Stone (2004: 557) speaks of consultants as 'representational intermediaries'. Knox (2008) has referred to the 'exchange professionals' involved with the enchantment and re-enchantment of built-environment forms.

3   The metrics used in world-city analysis are narrow. Other metrics such as debt, remittances and Islamic financial connections produce different maps of global connections.

References

Abbott, C. and Adler, S. (1989) 'Historical analysis as a planning tool', *Journal of the American Planning Association* 55: 467–73.

Abram, S. (2014) 'The time it takes: temporalities of planning', *Journal of the Royal Anthropological Institute* 20: 129–47.

Abram, S. and Weszkalays, G. (2011) 'Introduction: anthropologies of planning – temporality, imagination, and ethnography', *Journal of Global and Historical Anthropology* 61: 3–18.

Acuto, M. (2010) 'High-rise Dubai urban entrepreneurialism and the technology of symbolic power', *Cities* 27: 272–84.

Acuto, M. and Rayner, S. (2016) 'City networks: breaking gridlocks or forging (new) lock-ins?', *International Affairs* 92: 1147–66.

Addie, J. P. D. (2017) 'From the urban university to universities in urban society', *Regional Studies* 51: 1089–99.

Alexander, S. and Gleeson, B. (2019) *Degrowth in the Suburbs: A Radical Urban Imaginary*. Basingstoke: Palgrave Macmillan.

Ali, S. H. and Keil, R. (2006) 'Global cities and the spread of infectious disease: the case of severe acute respiratory syndrome (SARS) in Toronto, Canada', *Urban Studies* 43: 491–509.

Allen, J. (2005) *Lost Geographies of Power*. Oxford: Blackwell.

Allen, J. (2010) 'Powerful city networks: more than connections, less than domination and control', *Urban Studies* 47: 2895–911.

Allmendinger, P. and Haughton, G. (2009) 'Soft spaces, fuzzy boundaries, and metagovernance: the new spatial planning in the Thames Gateway', *Environment and Planning A* 41: 617–33.

Alvaredo, F., Chancel, L., Piketty, T., Saez, E. and Zucman, G. (2017) *World Inequality Report 2018*. Paris: World Inequality Lab.

Ambrose, P. (1994) *Urban Process and Power*. London: Routledge.

Angel, S. and Blei, A. M. (2016) 'The spatial structure of American cities: the great majority of workplaces are no longer in CBDs, employment sub-centers, or live-work communities', *Cities* 51: 21–35.

Angel, S. and Boonyabancha, S. (1988) 'Land sharing as an alternative to eviction', *Third World Planning Review* 10: 107–27.

Anzorena, J., et al. (1998) 'Reducing urban poverty: some lessons from experience', *Environment & Urbanization* 10: 167–86.

Aragón, P., Kaltenbrunner, A., Calleja-López, A., Pereira, A., Monterde, A., Barandiaran, X. E. and Gómez, V. (2017) 'Deliberative platform design: the case study of the online discussions in Decidim Barcelona', 277–87 in Ciampaglia, G. L., Mashhadi, A. and Yasseri, T. (eds.), *Social Informatics: 9th International Conference on Social Informatics Part II*. Cham: Springer.

Arias-Loyola, M. and Vergara-Perucich, F. (2021) 'Co-producing the right to fail: resilient grassroot cooperativism in a Chilean informal settlement', *International Development Planning Review* 43: 33–62.

Arnstein, S. R. (1969) 'A ladder of citizen participation', *Journal of the American Institute of Planners* 35: 216–24.

Augé, M. (1995) *Non-Places: Introduction to an Anthropology of Supermodernity*. London: Verso.

Baigent, E. (2004) 'Patrick Geddes, Lewis Mumford and Jean Gottmann: divisions over "megalopolis"', *Progress in Human Geography* 28: 687–700.

Bäing, A. S. and Webb, B. (2020) *Planning Through Zoning*. London: Royal Town Planning Institute.

Baldacchino, G. (2010) *Island Enclaves: Offshoring Strategies, Creative Governance, and Subnational Island Jurisdictions*. Montreal: McGill-Queen's University Press.

Balz, V. and Zonneveld, W. (2018) 'Transformations of planning rationales: changing spaces for governance in recent Dutch national planning', *Planning Theory & Practice* 19: 363–84.

Banham, R. (1971) *Los Angeles: The Architecture of Four Ecologies*. London: Allen Lane.

Banham, R., Barker, P., Hall, P. and Price, C. (1969) 'Non-plan: an experiment in freedom', *New Society* 13: 435–41.

Banister, J. and Anable, J. (2009) 'Transport policies and climate change', 55–69 in Davoudi, S., Crawford, J. and Mehmood, A. (eds.), *Planning for Climate Change: Strategies for Mitigation and Adaptation for Spatial Planners*. Abingdon: Earthscan.

Banks, S. and Carpenter, M. (2017) 'Researching the local politics and practices of radical Community Development Projects in 1970s Britain', *Community Development Journal* 52: 226–46.

Barnett, C. and Parnell, S. (2018) 'Spatial rationalities and the possibilities for planning in the New Urban Agenda for Sustainable Development', 25–36 in Bhan, G., Srinivas, S. and Watson, V. (eds.), *The Routledge Companion to Planning in the Global South*. Abingdon: Routledge.

Barton, H. (2009) 'Land use planning and health and well-being', *Land Use Policy* 26: S115–S123.

Batey, P. (2018) 'The history of planning methodology', 46–59 in Hein, C. (ed.), *The Routledge Handbook of Planning History*. Abingdon: Routledge.

Batty, M. (2018) *Inventing Future Cities.* Cambridge, MA: MIT Press.

Batty, M. and Marshall, S. J. (2012) 'The origins of complexity theory in cities and planning', 21–45 in Portugali, J., Meyer, H., Solk, E. and Tan, E. (eds.), *Complexity Theories of Cities Have Come of Age.* Dordrecht: Springer.

Bayly, C. A. (2000) *The Birth of the Modern World.* Cambridge: Cambridge University Press.

Beatley, T. (2012) 'Sustainability in planning: the arc and trajectory of a movement, and new directions for the twenty-first century', 91–124 in Sanyal, B., Vale, L. J. and Rosen, C. D. (eds.), *Planning Ideas that Matter: Livability, Territoriality, Governance and Reflective Practice.* Cambridge, MA: MIT Press.

Beatley, T. (2016) *Handbook of Biophilic City Planning and Design.* Washington, DC: Island Press.

Beauregard, R. (2006) *When America Became Suburban.* Minneapolis: University of Minnesota Press.

Beauregard, R. (2018) 'The entanglements of uncertainty', *Journal of Planning Education and Research,* 0739456X18783038.

Beck, U. (2002) *Risk Society: Toward a New Modernity.* London: Sage.

Beck, U., Bonss, W. and Lau, C. (2003) 'The theory of reflexive modernization: problematic, hypotheses and research program', *Theory, Culture & Society* 20: 1–33.

Beeckmans, L. and Lagae, J. (2015) 'Kinshasa's syndrome planning in historical perspective: from Belgian colonial capital to self-constructed megalopolis', 201–24 in Silva, C. N. (ed.), *Urban Planning in Sub-Saharan Africa: Colonial and Post-Colonial Planning Cultures.* Abingdon: Routledge.

Benjamin, S. (2008) 'Occupancy urbanism: radicalizing politics and economy beyond policy and programs', *International Journal of Urban and Regional Research* 32: 719–29.

Benjamin, S., Arifin, A. and Sarjana, F. (1985) 'The housing costs of low-income *kampung* dwellers: a study of product and process in Indonesian cities', *Habitat International* 9: 91–110.

Ben-Joseph, E. (2012) 'Codes and standards', 352–66 in Weber, R. and Crane, R. (eds.), *The Oxford Handbook of Urban Planning.* Oxford: Oxford University Press.

Berry, J. and McGreal, S. (1995) 'European cities: the interaction of planning systems, property markets and real estate investment', 1–16 in Berry, J. and McGreal, S. (eds.), *European Cities, Planning Systems and Property Markets.* London: Spon Press.

Bertman, S. (1998) *Hyperculture: The Human Costs of Speed.* Westport, CT: Praeger.

Bhan, G. (2019) 'Notes on a Southern urban practice', *Environment and Urbanization* 31: 639–54.

Bhan, G., Srinivas, S. and Watson, V. (2018) 'Introduction', 1–22 in Bhan, G.,

Srinivas, S. and Watson, V. (eds.), *The Routledge Companion to Planning in the Global South*. Abingdon: Routledge.

Birch, E. L. (1980) 'Radburn and the American planning movement: the persistence of an idea', *Journal of the American Planning Association* 46: 424–39.

Blackburn, S. and Marques, C. (2013) 'Mega-urbanization on the coast: global context and key trends in the twenty-first century', 1–21 in Pelling, M. and Blackburn, S. (eds.), *Megacities and the Coast: Risk, Resilience and Transformation*. Abingdon: Routledge.

Blockmans, W. P. (1989) 'Voracious states and obstructing cities: an aspect of state formation in preindustrial Europe', 218–50 in Tilly, C. and Blockmans, W. P. (eds.), *Cities and the Rise of States in Europe., A.D. 1000 to 1800*. Boulder, CO: Westview Press.

Blomley, N. (2014) 'Making space for property', *Annals of the Association of American Geographers* 104: 1291–1306.

Blomley, N. (2017) 'Land use, planning, and the "difficult character of property"', *Planning Theory & Practice* 18: 351–64.

Boli, J. (1999) 'Conclusion: world authority structures and legitimations', 267–300 in Boli, J. and Thomas, G. M. (eds.), *Constructing World Culture: International Nongovernmental Organizations since 1875*. Stanford, CA: Stanford University Press.

Boli, J. and Thomas, G. M. (1999a) 'Introduction', 1–10 in Boli, J. and Thomas, G. M. (eds.), *Constructing World Culture: International Nongovernmental Organizations since 1875*. Stanford, CA: Stanford University Press.

Boli, J. and Thomas, G. M. (1999b) 'INGOs and the organization of world culture', 13–49 in Boli, J. and Thomas, G. M. (eds.), *Constructing World Culture: International Nongovernmental Organizations since 1875*. Stanford, CA: Stanford University Press.

Bollens, S. A. (2002) 'Urban planning and intergroup conflict: confronting a fractured public interest', *Journal of the American Planning Association* 68(1): 22–42.

Booth, P. (2011) 'Culture, planning and path dependence: some reflections on the problems of comparison', *Town Planning Review* 82: 13–28.

Booth, P. (2012) 'The unearned increment: property and the capture of betterment value in Britain and France', 74–93 in Ingram, G. and Hong, Y-H. (eds.), *Land Value Capture and Land Policies*. Cambridge, MA: Lincoln Institute of Land Policy.

Booth, P. et al. (2020) *The Wrong Answers to the Wrong Questions*. London: Town and Country Planning Association.

Bourdieu, P. (1984) *Distinction: A Social Critique of the Judgement of Taste*. Cambridge, MA: Harvard University Press.

Braithwaite, J. and Drahos, P. (2000) *Global Business Regulation*. Cambridge: Cambridge University Press.

Brand, R. and Gaffikin, F. (2007) 'Collaborative planning in an uncollaborative world', *Planning Theory* 6: 282–313.

Brenner, N. (1998) 'Between fixity and motion: accumulation, territorial organization and the historical geography of spatial scales', *Environment and Planning D* 16: 459–81.

Brindley, T., Rydin, Y. and Stoker, G. (1996) *Remaking Planning: The Politics of Urban Change.* London: Routledge.

Brinkley, C. and Hoch, C. (2018) 'The ebb and flow of planning specializations', *Journal of Planning Education and Research.* Available at: https://doi.org/10.1177%2F0739456X18774119 .

Bruegmann, R. (2005) *Sprawl: A Compact History.* Chicago: University of Chicago Press.

Bruton, M. J. (1984) 'Introduction: general planning and physical planning', 11–30 in Bruton, M. J. (ed.), *The Spirit and Purpose of Planning*, 2nd edn. London: Hutchinson.

Buliung, R. N. (2011) 'Wired people in wired places: stories about machines and the geography of activity', *Annals of the Association of American Geographers* 101: 1365–81.

Bulkely, H. and Castan Broto, V. (2013) 'Government by experiment? Global cities and the governing of climate change', *Transactions of the Institute of British Geographers* 38: 361–75.

Bunge, W. (2011 [1971]) *Fitzgerald: Geography of a Revolution.* Athens: University of Georgia Press.

Bunnell, T. (2004). *Malaysia, Modernity and the Multimedia Super Corridor: A Critical Geography of Intelligent Landscapes.* London: Routledge.

Bunnell, T. (2015) 'Antecedent cities and inter-referencing effects: learning from and extending beyond critiques of neoliberalisation', *Urban Studies* 52: 1983–2000.

Bunnell, T. and Das, D. (2010) 'Urban pulse – a geography of serial seduction: urban policy transfer from Kuala Lumpur to Hyderabad', *Urban Geography* 31: 277–84.

Bunnell, T., Miller, M. A., Phelps, N. A. and Taylor, J. (2013) 'Urban development in a decentralized Indonesia: two success stories?', *Pacific Affairs* 86: 857–76.

Burawoy, M. (1998) 'The extended case method', *Sociological Theory* 16: 4–33.

Butler, T. and Lees, L. (2006) 'Super-gentrification in Barnsbury, London: globalization and gentrifying global elites at the neighbourhood level', *Transactions of the Institute of British Geographers* 31: 467–87.

Calthorpe, P. (2005) 'New urbanism: principles or style?', 16–38 in Fishman, R. (ed.), *New Urbanism: Peter Calthorpe vs. Lars Lerup.* Ann Arbor: University of Michigan.

Camaren, P. and Swilling, M. (2012) *Sustainable, Resource Efficient Cities: Making it Happen!* Nairobi: United Nations Environment Programme.

Canuto, M. A. et al. (2018) 'Ancient lowland Maya complexity as revealed by airborne laser scanning of northern Guatemala', *Science* 361(6409), eaau0137.

Cardoso, F. H. and Faletto, E. (1979) *Dependency and Development in Latin America*. Berkeley: University of California Press.

Carey, R., Larsen, K., Sheridan, J. and Candy, S. (2016) *Melbourne's Food Future: Planning a Resilient City Foodbowl*. Melbourne: VEIL.

Carmona, M. (2011) 'Design coding: mediating the tyrannies of practice', 54–73 in Tiesdell, S. and Adams, D. (eds.), *Urban Design in the Real Estate Development Process*. Oxford: Wiley-Blackwell.

Caro, R. A. (1974) *The Power Broker: Robert Moses and the Fall of New York*. New York: Alfred Knopf.

Carolini, G. (2015) 'Valuing possibility: south–south cooperation and participatory budgeting in Maputo, Mozambique', 266–84 in Silva, C. N. (ed.), *Urban Planning in Sub-Saharan Africa: Colonial and Post-Colonial Planning Cultures*. Abingdon: Routledge.

Casey, E. (1997) *The Fate of Place: A Philosophical History*. Berkeley: University of California Press.

Castells, M. (1983) *The City and the Grassroots*. London: Edward Arnold.

Castells, M. (2005) 'Space of flows, space of places: materials for a theory of urbanism in the information age', 45–63 in Sanyal, B. (ed.), *Comparative Planning Cultures*. Abingdon: Routledge.

CEC (Commission of the European Communities) (1997) *The EU Compendium of Spatial Planning Systems and Policies*. Luxembourg: CEC.

Charney, I. (2012) 'The real estate development industry', 722–38 in Weber, R. and Crane, R. (eds.), *The Oxford Handbook of Urban Planning*. Oxford: Oxford University Press.

Chatterton, P. (2002) '"Squatting is still legal, necessary and free": a brief intervention in the corporate city', *Antipode* 34: 1–7.

Chen, C. L. (2012) 'Reshaping Chinese space-economy through high-speed trains: opportunities and challenges', *Journal of Transport Geography* 22: 312–16.

Cherry, G. E. (1996) *Town Planning in Britain since 1900*. Oxford: Blackwell.

Clark, P. (2013) 'Introduction', 1–26 in Clark, P. (ed.), *The Handbook of Cities in World History*. Oxford: Oxford University Press.

Cleaver, F. (2004) 'The social embeddedness of agency and decision-making', 271–7 in Hickey, S. and Mohan, G. (eds.), *Participation: From Tyranny to Transformation*. London: Zed Books.

Colomb, C. (2017) 'Participation and conflict in the formation of neighbourhood areas and forums in "super-diverse" cities', 127–44 in Brownill, S. and Bradley, Q. (eds.), *Localism and Neighbourhood Planning: Power to the People?* Bristol: Policy Press.

Colomb, C. and Tomaney, J. (2016) 'Territorial politics, devolution and spatial planning in the UK: results, prospects, lessons', *Planning Practice & Research* 31: 1–22.

Communist Party of Australia (1970) *Plan for Melbourne*, vol. 1. Melbourne: Communist Party of Australia.

Cook, I. R., Ward, S. V. and Ward, K. (2014) 'A springtime journey to the Soviet Union: postwar planning and policy mobilities through the Iron Curtain', *International Journal of Urban and Regional Research* 38: 805–22.

Corburn, J. (2004) 'Confronting the challenges in reconnecting urban planning and public health', *American Journal of Public Health* 94: 541–6.

Corburn, J. (2012) 'Reconnecting urban planning and public health', 352–66 in Weber, R. and Crane, R. (eds.), *The Oxford Handbook of Urban Planning*. Oxford: Oxford University Press.

Corfield, P. J. (2013) 'Conclusion: cities in time', 828–46 in Clark, P. (ed.), *The Handbook of Cities in World History*. Oxford: Oxford University Press.

Cowen, D. (2014) *The Deadly Life of Logistics: Mapping Violence in Global Trade*. Minneapolis: University of Minnesota Press.

Cox, K. R. (1995) 'Globalisation, competition and the politics of local economic development', *Urban Studies* 32: 213–24.

Crook, T. (2016) 'Capturing development value through *de jure* national taxation: the English experience', 37–62 in Crook, T., Henneberry, J. and Whitehead, C. (eds.), *Planning Gain: Providing Infrastructure and Affordable Housing*. Chichester: Wiley.

Crook, T., Henneberry, J. and Whitehead, C. (2016) 'Introduction', 1–19 in Crook, T., Henneberry, J. and Whitehead, C. (eds.), *Planning Gain: Providing Infrastructure and Affordable Housing*. Chichester: Wiley.

Cullingworth, J. B. (1980) *Environmental Planning, 1939–1969. Vol. 4. Land Values, Compensation and Betterment*. Norwich: HMSO.

Curran, D. (2012) 'Rogue landlords and "beds in sheds"'. London: London Councils.

Currier, J. (2008) 'Art and power in the new China', *Town Planning Review* 79: 237–65.

Curtis, S. (2016) *Global Cities and Global Order*. Oxford: Oxford University Press.

Cwerner, S. (2009) 'Introducing aeromobilities', 1–21 in Cwerner, S., Kesselring, S. and Urry, J. (eds.), *Aeromobilities*. Abingdon: Routledge.

Davidoff, P. (1965) 'Advocacy and pluralism in planning', *Journal of the American Institute of Planners* 31: 331–8.

Davidson, K., Coenen, L., Acuto, M. and Gleeson, B. (2019) 'Reconfiguring urban governance in an age of rising city networks: a research agenda', *Urban Studies* 56: 3540–55.

Davies, H. W. E., Edwards, D., Hooper, J. A. and Punter, J. V. (1989) *Planning Control in Western Europe*. London: HMSO.

Davoudi, S., Crawford, J. and Mehmood, A. (2009) 'Climate change and spatial planning responses', 7–18 in Davoudi, S., Crawford, J. and Mehmood, A. (eds.), *Planning for Climate Change: Strategies for Mitigation and Adaptation for Spatial Planners*. Abingdon: Earthscan.

De Jong, J. K. (2013) *New Suburbanisms*. Abingdon: Routledge.

Depaulle, P. (1922) 'Functions of comparative law', *Harvard Law Review* 35: 838–58.

Dezalay, Y. and Garth, B. G. (1996) *Dealing in Virtue: International Commercial Arbitration and the Construction of a Transnational Legal Order*. Chicago: University of Chicago Press.

Dick, H. W. and Rimmer, P. J. (1998) 'Beyond the third world city: the new urban geography of South-East Asia', *Urban Studies* 35: 2303–21.

Dietz, T., Ostrom, E. and Stern, P. C. (2003) 'The struggle to govern the commons', *Science* 302(5652): 1907–12.

Dore, R. P. (1990) *British Factory: Japanese Factory*. Berkeley: University of California Press.

Dorling, D. (2011) *Injustices: Why Social Inequality Persists*. Bristol: Policy Press.

Douglass, M. (2016) 'Creative communities and the cultural economy – : Insadong, chaebol urbanism and the local state in Seoul', *Cities* 56: 148–55.

Dovey, K. (1985) 'Home and homelessness', 33–64 in Altman, I. and Werner, C. M. (eds.), *Home Environments*. New York: Springer.

Dovey, K. and King, R. (2011) 'Forms of informality: morphology and visibility of informal settlements', *Built Environment* 37: 11–29.

Doxiadis, C. A. (1962) 'Ecumenopolis: toward a universal city', *Ekistics* 13: 3–18.

Dühr, S., Colomb, C. and Nadin, V. (2010) *European Spatial Planning and Territorial Cooperation*. Abingdon: Routledge.

Dunford, M. (1996) 'Disparities in employment, productivity and output in the EU: the roles of labour market governance and welfare regimes', *Regional Studies* 30: 339–57.

Dunham-Jones, E. and Williamson, J. (2009) *Retrofitting Suburbia: Urban Design Solutions for Redesigning Suburbs*. Chichester: Wiley.

Dyson, P. (2012) 'Slum tourism: representing and interpreting "reality" in Dharavi, Mumbai', *Tourism Geographies* 14: 254–72.

Ekers, M., Hamel, P. and Keil, R. (2012) 'Governing suburbia: modalities and mechanisms of suburban governance', *Regional Studies* 46: 405–22.

Elden, S. (2005) 'Missing the point: globalization, deterritorialization and the space of the world', *Transactions of the Institute of British Geographers* 30: 8–19.

Elliott, P. (2018) 'Who are we? Using the census to understand Victorian planners', *Planning News* 44: 26–7.

Elsheshtawy, Y. (2009) *Dubai: Behind an Urban Spectacle*. Abingdon: Routledge.

Entrikin, J. N. (1991) *The Betweenness of Place*. Baltimore: Johns Hopkins University Press.

Ernstson, H., Lawhon, M. and Duminy, J. (2014) 'Conceptual vectors of African urbanism: "engaged theory-making" and "platforms of engagement"', *Regional Studies* 48: 1563–77.

Esping-Anderson, G. (1990). *The Three Worlds of Welfare Capitalism*. Cambridge: Polity.

ESPON (2018) *COMPASS – Comparative Analysis of Territorial Governance and Spatial Planning Systems in Europe: Applied Research 2016–2018 Final Report*. Luxembourg: European Spatial Planning Observation Network.

Evans, N. (2002) '*Machi-zukuri* as a new paradigm in Japanese urban planning: reality or myth', *Japan Forum* 14: 443–64.

Fainstein, S. (2010) *The Just City*. Ithaca, NY: Cornell University Press.

Fainstein, S. (2012) 'Land value capture and justice', 21–40 in Ingram, G. and Hong, Y.-H. (eds.), *Land Value Capture and Land Policies*. Cambridge, MA: Lincoln Institute of Land Policy.

Faludi, A. (2015) 'The European Union context of national planning', 259–95 in Knaap, G.-J., Nedović-Budić, Z. and Carbonell, A. (eds.), *Planning for States and Nation-States in the U.S. and Europe*. Cambridge, MA: Lincoln Institute of Land Policy.

Farson, R. (1996) *Management of the Absurd: Paradoxes in Leadership*. New York: Simon and Schuster.

Ferguson, J. (2006) *Global Shadows: Africa in the Neoliberal World Order*. Durham, NC: Duke University Press.

Fincher, R. and Iveson, K. (2008) *Planning and Diversity in the City*. Basingstoke: Palgrave Macmillan.

Fishman, R. (1987) *Bourgeois Utopias: The Rise and Fall of Suburbia*. Baltimore: Johns Hopkins University Press.

Florida, R., Gulden, T. and Mellander, C. (2008) 'The rise of the mega-region', *Cambridge Journal of Regions, Economy and Society* 1: 459–76.

Flyvbjerg, B. (1998) *Rationality and Power: Democracy in Practice*. Chicago: University of Chicago Press.

Flyvbjerg, B. (2006) 'Five misunderstandings about case-study research', *Qualitative Inquiry* 12: 219–45.

Flyvbjerg, B. and Sunstein, C. R. (2016) 'The principle of the malevolent hiding hand; or, the planning fallacy writ large', *Social Research: An International Quarterly* 83: 979–1004.

Folmer, E. and Risselada, A. (2013) 'Planning the neighbourhood economy: land-use plans and the economic potential of urban residential neighbourhoods in the Netherlands', *European Planning Studies* 21: 1873–94.

Forester, J. (1993) *Critical Theory, Public Policy, and Planning Practice*. Albany, NY: SUNY Press.

Forester, J. (1999) *The Deliberative Practitioner: Encouraging Participatory Planning Processes*. Cambridge, MA: MIT Press.

Forester, J. (2006) 'Policy analysis as critical listening', 124–51 in Moran, M., Rein, M. and Goodin, R. (eds.), *The Oxford Handbook of Public Policy*. Oxford: Oxford University Press.

Frank, A. I. and Silver, C. (2018) 'Envisaging the future of planning and planning education', 235–50 in Frank, A. I. and Silver, C. (eds.), *Urban Planning Education: Beginnings, Global Movement and Future Prospects*. Dordrecht: Springer.

Freestone, R., Randolph, B. and Pinnegar, S. (2018) 'Suburbanization in Australia', 72–86 in Hanlon, B. and Vicino, T. (eds.), *The Routledge Companion to the Suburbs*. Abingdon: Routledge.

Freestone, R., Williams, P. and Borden, A. (2006) 'Flybuy cities: some planning aspects of airport privatization in Australia', *Urban Policy and Research* 24: 491–508.

French, A. (1995) *Plans, Pragmatism and People: The Legacy of Soviet Planning for Today's Cities*. London: UCL Press.

Friedmann, J. (1987) *Planning in the Public Domain: From Knowledge to Action*. Princeton, NJ: Princeton University Press.

Friedmann, J. (2005) *China's Urban Transition*. Minneapolis: University of Minnesota Press.

Friedmann, J. (2006) 'Globalization and the emerging culture of planning', *Progress in Planning* 64: 183–234.

Friedmann, J. and Wolff, G. (1982) 'World city formation: an agenda for research and action', *International Journal of Urban and Regional Research* 6: 309–44.

Friend, J. K. and Hickling, A. (2005) *Planning Under Pressure: The Strategic Choice Approach*. Abingdon: Routledge.

Friend, J. K. and Jessop, W. N. (1969) *Local Government and Strategic Choice: An Operational Research*. London: Tavistock.

Frug, G. (2000) *City Making: The Building of Cities Without Walls*. Princeton, NJ: Princeton University Press.

Galster, G. (2019) 'Why shrinking cities are not mirror images of growing cities: a research agenda of six testable propositions', *Urban Affairs Review* 55: 355–72.

Gandy, M. (2014) *The Fabric of Space: Water, Modernity, and the Urban Imagination*. Cambridge, MA: MIT Press.

García Mejuto, D. (2017) 'A Europe of multiple flows: contested discursive integration in trans-European transport infrastructure policy-making', *European Urban and Regional Studies* 24: 425–41.

Garreau, J. (1991) *Edge City: Life on the New Frontier*. New York: Doubleday.

Garvin, A. (2009) 'Planners as leaders', 25–33 in Hack, G., Birth, E. L.,

Sedway, P. H. and Silver, J. (eds.), *Local Planning: Contemporary Problems and Prospects*. Washington, DC: ICMA Press.

Geddes, P. (1904) *Civics: As Applied Sociology*. Project Gutenberg.

Gibson, C., Klocker, N., Borger, E. and Kerr, S. M. (2018) 'Malleable homes and mutual possessions: caring and sharing in the extended family household as a resource for survival', 35–49 in Ince, A. and Hall, S. M. (eds.), *Sharing Economies in Times of Crisis*. Abingdon: Routledge.

Gibson-Graham, J. K., Cameron, J. and Healy, S. (2013) *Take Back the Economy: An Ethical Guide for Transforming our Communities*. Minneapolis: University of Minnesota Press.

Gilbert, A. (2007) 'The return of the slum: does language matter?', *International Journal of Urban and Regional Research* 31: 697–713.

Giles-Corti, B. et al. (2016) 'City planning and population health: a global challenge', *The Lancet* 388: 2912–24.

Gillette, H., Jr (2011) *Between Justice and Beauty: Race, Planning, and the Failure of Urban Policy in Washington*. Philadelphia: University of Pennsylvania Press.

Ginkel, H. J. A. and Marcotullio, P. J. (2005) 'Asian urbanization and local and global environmental challenges', 11–35 in Keiner, M., Koll-Schretzenmayr, M. and Schmid, W. A. (eds.), *Managing Urban Futures: Sustainability and Urban Growth in Developing Countries*. Aldershot: Ashgate.

Glaeser, E. (2011) *The Triumph of the City*. London: Macmillan.

Glasson, J. (1978) *An Introduction to Regional Planning*, 2nd edn. London: Hutchinson.

Gleeson, B. (2006) *Australian Heartlands: Making Space for Hope in the Suburbs*. Sydney: Allen & Unwin.

Gleeson, B. and Low, N. (2000) *Australian Urban Planning: New Challenges, New Agendas*. Sydney: Allen & Unwin.

Gold, J. R. (1997) *The Experience of Modernism: Modern Architects and the Future of the City, 1928–53*. Abingdon: Routledge.

Goldman, M. (2011) 'Speculative urbanism and the making of the next world city', *International Journal of Urban and Regional Research* 35: 555–81.

Gombay, N. (2018) 'Just enough to survive: economic citizenship in the context of indigenous land claims', 160–74 in Ince, A. and Hall, S. M. (eds.), *Sharing Economies in Times of Crisis*. Abingdon: Routledge.

Gomez-Baggethun, E. and Barton, D. N. (2013) 'Classifying and valuing ecosystems services for urban planning', *Ecological Economics* 86: 235–45.

Gottlieb, R. (2007) *Reinventing Los Angeles: Nature and Community in the Global City*. Cambridge, MA: MIT Press.

Gottmann, J. (1961) *Megalopolis*. Cambridge, MA: MIT Press.

Graham, S. and Marvin, S. (2000) *Splintering Urbanism*. Abingdon: Routledge.

Grant, R. (2005) 'The emergence of gated communities in a West African context: evidence from Greater Accra, Ghana', *Urban Geography* 26: 661–83.

Gray, K. and Gills, B. K. (2016) 'South–south cooperation and the rise of the Global South', *Third World Quarterly* 37: 557–74.

Green, N. and Handley, J. (2009) 'Patterns of settlement compared', 46–54 in Davoudi, S., Crawford, J. and Mehmood, A. (eds.), *Planning for Climate Change: Strategies for Mitigation and Adaptation for Spatial Planners*. Abingdon: Earthscan.

Hack, G. (2018) *Site Planning: International Practice*. Cambridge, MA: MIT Press.

Hall, P. (1977) 'The inner cities dilemma', *New Society* 39: 223–5.

Hall, P. (1980) *Great Planning Disasters*. London: Weidenfeld and Nicolson.

Hall, P. and Tewdwr-Jones, M. (2020) *Urban and Regional Planning*. Abingdon: Routledge.

Hall, S. M. and Ince, A. (2018) 'Introduction: sharing economies in times of crisis', 1–15 in Ince, A. and Hall, S. M. (eds.), *Sharing Economies in Times of Crisis*. Abingdon: Routledge.

Halsnæs, K. and Laursen, N. V. (2009) 'Climate change vulnerability: a new threat to poverty alleviation in developing countries', 83–93 in Davoudi, S., Crawford, J. and Mehmood, A. (eds.), *Planning for Climate Change: Strategies for Mitigation and Adaptation for Spatial Planners*. Abingdon: Earthscan.

Hanes, J. E. (1993) 'From megalopolis to megaroporisu', *Journal of Urban History* 19: 56–94.

Hanson, R. (2017) *Suburb: Planning Politics and the Public Interest*. Ithaca, NY: Cornell University Press.

Hardin, G. (1968) 'The tragedy of the commons', *Science* 162: 1243–8.

Harding, A. (1997) 'Urban regimes in a Europe of the cities?', *European Urban and Regional Studies* 4: 291–314.

Hardy, D. and Ward, C. (1984) *Arcadia for All: The Legacy of the Makeshift Landscape*. London: Mansell.

Harris, J. and Tewdwr-Jones, M. (2010) 'Valuing ecosystem services in planning', *Town and Country Planning* 79: 228.

Harris, R. (2004) *Creeping Conformity: How Canada Became Suburban, 1900–1960*. Toronto: University of Toronto Press.

Harrison, P. (2015) 'South–south relationships and the transfer of "best practice": the case of Johannesburg, South Africa', *International Development Planning Review* 37: 205–33.

Harrison, P., Rubin, M., Appelbaum, A. and Dittgen, R. (2019) 'Corridors of freedom: analyzing Johannesburg's ambitious inclusionary transit-oriented development', *Journal of Planning Education and Research* 39: 456–68.

Harvey, D. (1974) 'What kind of geography for what kind of public policy?', *Transactions of the Institute of British Geographers* 63: 18–24.

Harvey, D. (1985) *The Urbanization of Capital*. Oxford: Blackwell.

Harvey, D. (2000) *Spaces of Hope*. Edinburgh: Edinburgh University Press.

Hausman, D. M. and Welch, B. (2010) 'Debate: to nudge or not to nudge', *Journal of Political Philosophy* 18: 123–36.

Hayden, D. (1980) 'What would a non-sexist city be like? Speculations on housing, urban design, and human work', *Signs: Journal of Women in Culture and Society* 5(S3): S170–S187.

Hayden, D. (1997) *The Power of Place: Urban Landscapes as Public History*. Cambridge, MA: MIT Press.

Healey, P. (1997) *Collaborative Planning: Shaping Places in Fragmented Societies*. Basingstoke: Palgrave Macmillan.

Healey, P. (2004) 'The treatment of space and place in the new strategic spatial planning in Europe', *International Journal of Urban and Regional Research* 28: 45–67.

Healey, P. (2007) *Urban Complexity and Spatial Strategies: Towards a Relational Planning for Our Times*. Abingdon: Routledge.

Healey, P. (2012) 'The universal and the contingent: some reflections on the transnational flow of planning ideas and practices', *Planning Theory* 11: 188–207.

Healey, P., McNamara, P., Elson, M. and Doak, A. (1988) *Land Use Planning and the Mediation of Urban Change: The British Planning System in Practice*. Cambridge: Cambridge University Press.

Healey, P. and Williams, R. (1993) 'European urban planning systems: diversity and convergence', *Urban Studies* 30: 701–20.

Heidegger, M. (2010) *Being and Time*. Albany, NY: SUNY Press.

Hein, C. (2018) 'The what, why, and how of planning history', 1–10 in Hein, C. (ed.), *The Routledge Handbook of Planning History*. Abingdon: Routledge.

Hickey, S. and Mohan, G. (2004) 'Towards participation as transformation: critical themes and challenges', 3–24 in Hickey, S. and Mohan, G. (eds.), *Participation: From Tyranny to Transformation*. London: Zed Books.

Hirt, S. (2007) 'The devil is in the definitions: contrasting American and German approaches to zoning', *Journal of the American Planning Association* 73: 436–50.

Hoch, C. (2019) *Pragmatic Spatial Planning: Practical Theory for Professionals*. Abingdon: Routledge.

Holston, J. (2009) *Insurgent Citizenship: Disjunctions of Democracy and Modernity in Brazil*. Princeton, NJ: Princeton University Press.

Holston, J. and Appadurai, A. (1999) 'Introduction', in Holston, J. and Appadurai, A. (eds.), *Cities and Citizenship*. Durham, NC: Duke University Press.

Home, R. (2013) *Of Planting and Planning: The Making of British Colonial Cities*, 2nd edn. Abingdon: Routledge.

Home, R. (2015) 'Colonial urban planning in Anglophone Africa', 53–66 in Silva, C. N. (ed.), *Urban Planning in Sub-Saharan Africa: Colonial and Post-Colonial Planning Cultures*. Abingdon: Routledge.

Home, R. (2018) 'Global systems foundations of the discipline', 91–106 in Hein, C. (ed.) *The Routledge Handbook of Planning History*. Abingdon: Routledge.

Hopkins, L. D. (2001) *Urban Development: The Logic of Making Plans*. Washington, DC: Island Press.

Hopkins, L. D. and Zapata, M. (2007) 'Engaging the future: tools for effective planning practice', 1–18 in Hopkins, L. D. and Zapata, M. (eds.), *Engaging the Future: Forecasts, Scenarios, Plans, and Projects*. Cambridge, MA: Lincoln Institute of Land Policy.

Howard, E. (1902) *Garden Cities of Tomorrow*. London: S. Sonnenschein.

Imrie, R. F. (1996) *Disability and the City: International Perspectives*. London: Sage.

Ingram, G. and Hong, Y.-H. (2012) 'Land value capture: types and outcomes', 3–18 in Ingram, G. and Hong, Y.-H. (eds.), *Land Value Capture and Land Policies*. Cambridge, MA: Lincoln Institute of Land Policy.

Innes, J. E. and Booher, D. E. (2010) *Planning with Complexity: An Introduction to Collaborative Rationality for Public Policy*. Abingdon: Routledge.

Institute for Sustainable Communities (2015) *From Blueprint to Participatory Planning: Approaches to Sustainable Communities and Urbanization in China*. Beijing: Institute for Sustainable Communities.

Iveson, K. (2013) 'Cities within the city: do-it-yourself urbanism and the right to the city', *International Journal of Urban and Regional Research* 37: 941–56.

Jackson, S., Porter, L. and Johnson, L. C. (2017) *Planning in Indigenous Australia: From Imperial Foundations to Postcolonial Futures*. Abingdon: Routledge.

Jacobs, J. (1961) *The Death and Life of Great American Cities*. New York: Random House.

Jacobs, J. (1969) *The Economy of Cities*. London: Jonathan Cape.

Jacobs, J. M. (2012) 'Commentary: comparing comparative urbanisms', *Urban Geography* 33: 904–14.

Jensen, O. B. and Richardson, T. (2004) *Making European Space: Mobility, Power and Territorial Identity*. London: Routledge.

Jones, G. (2005) *Multinationals and Global Capitalism: From the Nineteenth to the Twenty-First Century*. Oxford: Oxford University Press.

Kanna, A. (2011) *Dubai: The City as Corporation*. Minneapolis: University of Minnesota Press.

Kasarda, J. D. and Lindsay, G. (2011) *Aerotropolis: The Way We'll Live Next*. New York: Farrar, Straus and Giroux.

Kayden, J. S. (2000) *Privately Owned Public Space: The New York City Experience*. New York: Wiley.

Keating, M. (1997) 'The invention of regions: political restructuring and territorial government in Western Europe', *Environment and Planning C* 15: 383–98.

Keil, R. and Macdonald, S. (2016) 'Rethinking urban political ecology from the outside in: greenbelts and boundaries in the post-suburban city', *Local Environment* 21: 1516–33.

Kellerman, A. (2012) *Personal Mobilities*. Abingdon: Routledge.

Kemeny, J. (1994) 'Understanding European rental systems', *SAUS Working Paper* 120. Bristol: University of Bristol.

Kenyon, S. and Lyons, G. (2007) 'Introducing multitasking to the study of travel and ICT: examining its extent and assessing its potential importance', *Transportation Research Part A: Policy and Practice* 41: 161–75.

Khanna, T. and Macomber, J. (2015) 'Government and the minimalist platform: business at the Kumbh Mela', 334–55 in Khanna, T., Macomber, J. and Chaturvedi, S., *Kumbh Mela: India's Pop-Up Mega-City*. Cambridge, MA: Harvard University Press.

Kim, H., Miao, J. T. and Phelps, N. A. (2021) 'International urban development leadership: Singapore, China and South Korea compared', 121–34 in Park, S. H., Shin, H. B. and Kang, H. S. (eds.), *Exporting Urban Korea? Reconsidering the Korean Urban Development Experience*. Abingdon: Routledge.

King, A. D. (1984) *The Bungalow: The Production of a Global Culture*. London: Routledge & Kegan Paul.

King, A. D. (2004) *Spaces of Global Cultures: Architecture, Urbanism, Identity*. Abingdon: Routledge.

Kitchin, R. and Dodge, M. (2011) *Code/Space: Software and Everyday Life*. Abingdon: Routledge.

Knaap, G.-J., Nedović-Budić, Z. and Carbonell, A. (2015a) 'Introduction', 1–26 in Knaap, G.-J., Nedović-Budić, Z. and Carbonell, A. (eds.), *Planning for States and Nation-States in the U.S. and Europe*. Cambridge, MA: Lincoln Institute of Land Policy.

Knaap, G.-J., Nedović-Budić, Z. and Carbonell, A. (2015b) 'Conclusion', 503–16 in Knaap, G.-J., Nedović-Budić, Z. and Carbonell, A. (eds.), *Planning for States and Nation-States in the U.S. and Europe*. Cambridge, MA: Lincoln Institute of Land Policy.

Knorringa, P., Peša, I., Leliveld, A. and Van Beers, C. (2016) 'Frugal innovation and development: aides or adversaries?', *The European Journal of Development Research* 28: 143–53.

Knox, P. L. (2008) *Metroburbia USA*. New Brunswick, NJ: Rutgers University Press.

Kolb, D. (2008) *Sprawling Places*. Athens: University of Georgia Press.

Kostof, S. (1991) *The City Shaped: Urban Patterns and Meanings Through History*. London: Thames & Hudson.

Kotkin, S. (1995) *Magnetic Mountain: Stalinism as a Civilization*. Berkeley: University of Chicago Press.

Kress, C. (2018) 'The German traditions of *städtebau* and *stadtlandschaft* and their diffusion through global exchange', 173–91 in Hein, C. (ed.) *The Routledge Handbook of Planning History*. Abingdon: Routledge.

Krumholz, N. (1982) 'A retrospective view of equity planning in Cleveland, 1969–1979', *Journal of the American Planning Association* 48: 163–74.

Kunzmann, K. R. (1996) 'Euro-megalopolis or theme park Europe? Scenarios for European spatial development', *International Planning Studies* 1: 143–63.

Kunzmann, K. R. (2005) 'Urban planning in the North: blueprint for the South?', 235–45 in Keiner, M., Koll-Schretzenmayr, M. and Schmid, W. A. (eds.), *Managing Urban Futures: Sustainability and Urban Growth in Developing Countries*. Aldershot: Ashgate.

Lake, A. A. and Townshend, T. (2006), 'Obesogenic environments: exploring the built and food environments', *Journal of the Royal Society for the Promotion of Health* 126: 262–7.

Larner, W. and Laurie, N. (2010) 'Travelling technocrats, embodied knowledges: globalising privatisation in telecoms and water', *Geoforum* 41: 218–26.

Larsson, G. (1997) 'Land readjustment: a tool for urban development', *Habitat International* 21: 141–52.

Latham, R. (2001) 'Identifying the contours of transboundary political life', 69–92 in Callaghy, T., Kassimir, R. and Latham, R. (eds.), *Intervention and Transnationalism in Africa: Global–Local Networks of Power*. Cambridge: Cambridge University Press.

Lauermann, J. (2018) 'Municipal statecraft: revisiting the geographies of the entrepreneurial city', *Progress in Human Geography* 42: 205–24.

Laurence, R. (2013) 'Planning and environment', 197–219 in Clark, P. (ed.), *The Handbook of Cities in World History*. Oxford: Oxford University Press.

Lawhon, M., Le Roux, L., Makina, A., Nsangi, G., Singh, A. and Sseviiri, H. (2020) 'Beyond southern urbanism? Imagining an urban geography of a world of cities', *Urban Geography*. Available at https://doi.org/10.1080/027 23638.2020.1734346.

Leach, R. (1997) 'Incrementalism and rationalism in local government reorganisation, 1957–1996', *Public Policy and Administration* 12: 59–72.

Lee, K. Y. (2011) *From Third World to First: The Singapore Story, 1965–2000*. New York: HarperCollins.

Leontidou, L. (1990) *The Mediterranean City in Transition: Social Change and Urban Development*. Cambridge: Cambridge University Press.

Levi-Faur, D. (2005) 'The global diffusion of regulatory capitalism', *Annals of the American Academy of Political and Social Science* 598: 12–32.

Levy, J. M. (2016) *Contemporary Urban Planning*, 11th edn. Abingdon: Routledge.

Li, Y. and Phelps, N. (2018) 'Megalopolis unbound: knowledge collaboration and functional polycentricity within and beyond the Yangtze River Delta Region in China, 2014', *Urban Studies* 55: 443–60.

Lindblom, C. E. (1959) 'The science of "muddling through"', *Public Administration Review* 19: 79–88.

Lindgren, M. and Bandhold, H. (2009) *Scenario Planning: The Link between Future and Strategy*, 2nd edn. Basingstoke: Palgrave Macmillan.

Lingua, V. and Servillo, L. (2014) 'The modernization of the Italian planning system', 127–43 in Reimer, M., Getimis, P. and Blotevogel, H. H. (eds.), *Spatial Planning Systems and Practices in Europe: A Comparative Perspective on Continuity and Changes*. Abingdon: Routledge.

Liu, C. (2008) 'Policy sprawl: the internal logic of spatial production', 286–337 in Mars, N. and Hornsby, A. (eds.), *The Chinese Dream: A Society Under Construction*. Rotterdam: 010 Publishers.

McCann, E. (2011) 'Urban policy mobilities and global circuits of knowledge: toward a research agenda', *Annals of the Association of American Geographers* 101(1): 107–30.

MacFarlane, C., Silver, J. and Truelove, Y. (2017) 'Cities within cities: intra-urban comparison of infrastructure in Mumbai, Delhi and Cape Town', *Urban Geography* 38: 1393–1417.

McGee, T. G. (1991) 'The emergence of desakota regions in Asia: expanding a hypothesis', 3–26 in Ginsburg, N., Koppel, B. and McGee, T. (eds.), *The Extended Metropolis: Settlement Transition in Asia*. Honolulu: University of Hawai'i Press.

McGlynn, S. (1993) 'Reviewing the rhetoric', 3–9 in Hayward, R. and McGlynn, S. (eds.), *Making Better Places: Urban Design Now*. Oxford: Butterworth-Heinemann.

McHarg, I. L. (1969) *Design with Nature*. New York: American Museum of Natural History Press.

MacKaye, B. (1962 [1928]) *The New Exploration: A Philosophy of Regional Planning*. Urbana, IL: University of Illinois Press.

McKenna, C. D. (2010) *The World's Newest Profession: Management Consulting in the Twentieth Century*. Cambridge: Cambridge University Press.

McLoughlin, J. B. (1969) *Urban Planning: A Systems Approach*. London: Faber.

Madanipour, A. (2017) *Cities in Time: Temporary Urbanism and the Future of the City*. London: Bloomsbury.

Magnusson, W. (2011) *Politics of Urbanism: Seeing Like a City*. Abingdon: Routledge.

Maier, K. (2012) 'Europeanization and changing planning in East-Central Europe: an Easterner's view', *Planning Practice and Research* 27: 137–54.

Majone, G. (1989) *Evidence, Argument, and Persuasion in the Policy Process.* New Haven, CT: Yale University Press.

March, A., Hurlimann, A. and Robins, J. (2013) 'Accreditation of Australian urban planners: building knowledge and competence', *Australian Planner* 50: 233–43.

March, A., Kornakova, M. and Leon, J. (2017) 'Integration and collective action: studies of urban planning and recovery after disasters', 1–12 in March, A. and Kornakova, M. (eds.), *Urban Planning for Disaster Recovery.* London: Butterworth-Heinemann.

Marcuse, P. (2012) 'Justice', 141–65 in Weber, R. and Crane, R. (eds.), *The Oxford Handbook of Urban Planning.* Oxford: Oxford University Press.

Marks, G., Hooghe, L. and Blank, K. (1996) 'European integration from the 1980s: state-centric v. multi-level governance', *Journal of Common Market Studies* 34: 341–78.

Marmot, A., Friel, S., Bell, R., Houweling, T. J. and Taylor, S. (2008) 'Closing the gap in a generation: health equity through action on the social determinants of health', *The Lancet* 372: 1661–9.

Mars, N. (2008) 'Cities without history', 520–37 in Mars, N. and Hornsby, A. (eds.), *The Chinese Dream: A Society under Construction.* Rotterdam: 010 Publishers.

Massey, D. B. (1989) *Space, Place and Gender.* Cambridge: Polity.

Meadows, D. H., Meadows, D. L., Randers, J. and Behrens, W. W. (1972) *The Limits to Growth.* New York: Potomac Books.

Meagher, K. (2018) 'Cannibalizing the informal economy: frugal innovation and economic inclusion in Africa', *The European Journal of Development Research* 30: 17–33.

Meller, H. (2005) *Patrick Geddes: Social Evolutionist and City Planner.* Abingdon: Routledge.

Meneses-Reyes, R. and Caballero-Juárez, J. A. (2014) 'The right to work on the street: public space and constitutional rights', *Planning Theory* 13: 370–86.

Meyer, J. W., Boli, J., Thomas, G. M. and Ramirez, F. O. (1997) 'World society and the nation-state', *American Journal of Sociology* 103: 144–81.

Miao, J. T. (2018a) 'Knowledge economy challenges for the post-developmental state: Tsukuba Science City as an in-between place', *Town Planning Review* 89: 61–84.

Miao, J. T. (2018b) 'Parallelism and evolution in transnational policy transfer networks: the case of Sino-Singapore Suzhou Industrial Park (SIP)', *Regional Studies* 52: 1191–1200.

Miao, J. T. and Maclennan, D. (2019) 'The rhetoric–reality gap of cities' success: learning from the practice of Scottish cities', *Regional Studies* 53: 1761–71.

Miao, J. T. and Phelps, N. A. (2019) 'The intrapreneurial state: Singapore's emergence in the smart and sustainable urban solutions field', *Territory, Politics, Governance* 7: 316–35.

Miao, J. T., Phelps, N. A., Lu, T. and Wang, C. C. (2019) 'The trials of China's technoburbia: the case of the Future Sci-Tech City Corridor in Hangzhou', *Urban Geography* 40: 1443–66.

Miller, D. (2010) *Stuff.* Cambridge: Polity.

Mills, C. W. (1959) *The Sociological Imagination.* Oxford: Oxford University Press.

Ministry of Housing, Communities and Local Government (2019) *Local Planning Authority Green Belt: England 2018/2019.* Available at: https:// assets.publishing.service.gov.uk/government/uploads/system/uploads/ attachment_data/file/856100/Green_Belt_Statistics_England_2018–19.pdf.

Ministry of Housing, Communities and Local Government (2020) *Planning for the Future.* White Paper. London: Ministry of Housing, Communities and Local Government.

Miraftab, F. (2009) 'Insurgent planning: situating radical planning in the global south', *Planning Theory* 8: 32–50.

Mitlin, D. (2008) 'Within and beyond the state: co-production as a route to political influence, power and transformation for grassroots organizations', *Environment and Urbanization* 20: 339–60.

Molotch, H. (1976) 'The city as a growth machine: toward a political economy of place', *American Journal of Sociology* 82: 309–32.

Molotch, H. (2004) *Where Stuff Comes From: How Toasters, Toilets, Cars, Computers and Many Other Things Come To Be As They Are.* Abingdon: Routledge.

Morgan, K. (2009) 'Feeding the city: the challenge of urban food planning', *International Planning Studies* 14: 341–8.

Morris, A. E. J. (1994) *History of Urban Form: Before the Industrial Revolution.* London: Longman.

Moser, S. (2018) 'Forest City, Malaysia, and Chinese expansionism', *Urban Geography* 39: 935–43.

Mouat, C., Legacy, C. and March, A. (2013) 'The problem is the solution: testing agonistic theory's potential to recast intractable planning disputes', *Urban Policy and Research* 31: 150–66.

Mozingo, L. A. (2016) *Pastoral Capitalism: A History of Suburban Corporate Landscapes.* Cambridge, MA: MIT Press.

Murray, M. J. (2017) *The Urbanism of Exception.* Cambridge: Cambridge University Press.

Nadin, V. and Stead, D. (2008) 'European spatial planning systems, social models and learning', *disP: The Planning Review* 44: 35–47.

Nadin, V. and Stead, D. (2013) 'Opening up the compendium: an evaluation of international comparative planning research methodologies', *European Planning Studies* 21: 1542–61.

Nasr, J. and Volait, M. (2003) 'Introduction: transporting planning', xi–xxxviii in Nasr, J. and Volait, M. (eds.), *Urbanism: Imported or Exported?* Chichester: Wiley.

Nasser, N. (2003) 'Planning for urban heritage places: reconciling conservation, tourism, and sustainable development', *Journal of Planning Literature* 17: 467–79.

Nathan, M. and Vandore, E. (2014) 'Here be startups: exploring London's "Tech City" digital cluster', *Environment and Planning A* 46: 2283–99.

Needham, B. (2006) *Planning, Law and Economics: The Rules We Make for Using Land.* Abingdon: Routledge.

Nelson, A. and Lang, R. (2011) *Megapolitan America: A New Vision for Understanding America's Metropolitan Geography.* Chicago: American Planning Association.

Neuman, P. (2005) 'The compact city fallacy', *Journal of Planning Education and Research* 25: 11–26.

Newman, P. and Thornley, A. (1996) *Urban Planning in Europe: International Competition, National Systems, and Planning Projects.* London: Routledge.

Newman, P. and Thornley, A. (2011) *Planning World Cities: Globalization and Urban Politics.* Basingstoke: Palgrave Macmillan.

Njoh, A. (2006) *Tradition, Culture and Development in Africa: Historical Lessons for Modern Development Planning.* Abingdon: Routledge.

Nolan, P. (1995) *China's Rise, Russia's Fall: Politics, Economics and Planning in the Transition from Stalinism.* Basingstoke: Palgrave Macmillan.

Nuissl, H. and Rink, D. (2005) 'The "production" of urban sprawl in eastern Germany as a phenomenon of post-socialist transformation', *Cities* 22: 123–34.

Nye, J. S. (1990) 'Soft power', *Foreign Policy* 80: 153–71.

Odendaal, N., Duminy, J. and Inkoom, D. K. B. (2015) 'The developmentalist origins and evolution of planning education in Sub-Saharan Africa c. 1940 to 2010', 285–99 in Silva, C. N. (ed.), *Urban Planning in Sub-Saharan Africa: Colonial and Post-Colonial Planning Cultures.* Abingdon: Routledge.

OECD (2017a) *Land-Use Planning Systems in the OECD: Country Fact Sheets.* Paris: OECD.

OECD (2017b) *The Governance of Land Use in the OECD Countries.* Paris: OECD.

Ohashi, H. and Phelps, N. A. (2020). 'Diversity in decline: the changing suburban fortunes of Tokyo Metropolis', *Cities* 103, 102693.

Oldfield, S. and Greyling, S. (2015) 'Waiting for the state: a politics of housing in South Africa', *Environment and Planning A* 47: 1100–12.

Olds, K. (2002) *Globalization and Urban Change: Capital, Culture, and Pacific Rim Mega-Projects.* Oxford: Oxford University Press.

Olowu, D. (2018) 'African urbanisation and democratisation: public policy, planning and public administration dilemmas', 59–69 in Bhan, G., Srinivas, S. and Watson, V. (eds.), *The Routledge Companion to Planning in the Global South*. Abingdon: Routledge.

Osterhammel, J. (2015) *The Transformation of the World: A Global History of the Nineteenth Century*. Princeton, NJ: Princeton University Press.

Othengrafen, F. and Reimer, M. (2013) 'The embeddedness of planning in cultural contexts: theoretical foundations for the analysis of dynamic planning cultures', *Environment and Planning A* 45: 1269–84.

Owens, S. and Cowell, R. (2011) *Land and Limits: Interpreting Sustainability in the Planning Process*. Abingdon: Routledge.

Parnell, S. (2018) 'Africa's urban planning palimpsest', 288–97 in Hein, C. (ed.), *The Routledge Handbook of Planning History*. Abingdon: Routledge.

Parnell, S., Pieterse, E. and Watson, V. (2009) 'Planning for cities in the global south: an African research agenda for sustainable human settlement', *Progress in Planning* 72: 233–41.

Passell, A. (2013) *Building the New Urbanism: Places, Professions, and Profits in the American Metropolitan Landscape*. Abingdon: Routledge.

Peck, J. and Theodore, N. (2015) *Fast Policy*. Minneapolis: University of Minnesota Press.

Pelling, M. and Blackburn, S. (2013) 'Case studies: governing social and environmental transformation in coastal megacities', 200–35 in Pelling, M. and Blackburn, S. (eds.), *Megacities and the Coast: Risk, Resilience and Transformation*. Abingdon: Routledge.

Perin, C. (1977) *Everything in its Place: Social Order and Land Use in America*. Princeton, NJ: Princeton University Press.

Perry, C. (2011 [1929]) '"The neighbourhood unit" from *The Regional Plan for New York and Its Environs*', 486–98 in Le Gates, R. T. and Stout, F. (eds.), *The City Reader*, 5th edn. Abingdon: Routledge.

Phelps, N. A. (2004) 'Clusters, dispersion and the spaces in between: for an economic geography of the banal', *Urban Studies* 41: 971–89.

Phelps, N. A. (2012a) *An Anatomy of Sprawl: Planning and Politics in Britain*. Abingdon: Routledge.

Phelps, N. A. (2012b) 'The growth machine stops? Urban politics and the making and remaking of an edge city', *Urban Affairs Review* 48: 670–700.

Phelps, N. A. (2015) *Sequel to Suburbia: Glimpses of America's Post-Suburban Future*. Cambridge, MA: MIT Press.

Phelps, N. A. (2017) *Interplaces: An Economic Geography of the Inter-Urban and International Economies*. Oxford: Oxford University Press.

Phelps, N. A., Bunnell, T., Miller, M. A. and Taylor, J. (2014) 'Urban inter-referencing within and beyond a decentralized Indonesia', *Cities* 39: 37–49.

Phelps, N. A., McNeill, D. and Parsons, N. (2002) 'In search of a European edge urban identity: trans-European networking among edge urban municipalities', *European Urban and Regional Studies* 9(3): 211–24.

Phelps, N. A., Parsons, N., Ballas, D. and Dowling, A. (2006) *Post-Suburban Europe: Planning and Politics at the Margins of Europe's Capital Cities*. Basingstoke: Palgrave Macmillan.

Phelps, N. and Tarazona Vento, A. (2015) 'Suburban governance in Western Europe', 155–76 in Hamel, P. and Keil, R. (eds.), *Suburban Governance: A Global View*. Toronto: University of Toronto Press.

Phelps, N. A. and Tewdwr-Jones, M. (2000) 'Scratching the surface of collaborative and associative governance: identifying the diversity of social action in institutional capacity building', *Environment and Planning A* 32: 111–30.

Phelps, N. A. and Tewdwr-Jones, M. (2008) 'If geography is anything, maybe it's planning's alter ego? Reflections on policy relevance in two disciplines concerned with place and space', *Transactions of the Institute of British Geographers* 33: 566–84.

Phelps, N. A. and Tewdwr-Jones, M. (2014) 'A man for all regions: Peter Hall and regional studies', *Regional Studies* 48: 1579–86.

Phelps, N. A. and Valler, D. (2018) 'Urban development and the politics of dissonance', *Territory, Politics, Governance* 6: 81–103.

Pickvance, C. (1982) 'Physical planning and market forces in urban development', 69–94 in Paris, C. (ed.), *Critical Readings in Planning Theory*. Oxford: Pergamon.

Pierre, J. (2005) 'Comparative urban governance: uncovering complex commonalities', *Urban Affairs Review* 40: 446–62.

Pieterse, E. (2008) *City Futures: Confronting the Crisis of Urban Development*. London: Zed Books.

Piketty, T. (2014) *Capital in the Twenty-First Century*. Cambridge, MA: Harvard University Press.

Pizarro, R. (2009) 'Urban form and climate change: towards appropriate development patterns to mitigate and adapt to global warming', 33–45 in Davoudi, S., Crawford, J. and Mehmood, A. (eds.), *Planning for Climate Change: Strategies for Mitigation and Adaptation for Spatial Planners*. Abingdon: Earthscan.

Pløger, J. (2004) 'Strife: urban planning and agonism', *Planning Theory* 3: 71–92.

Pollitt, C. (2000) 'Institutional amnesia: a paradox of the "information age"?', *Prometheus* 18: 5–16.

Pollock, L. S. (2009) 'Planners as private consultants', 188–9 in Hack, G., Birth, E. L., Sedway, P. H. and Silver, J. (eds.), *Local Planning: Contemporary Problems and Prospects*. Washington, DC: ICMA Press.

Potts, D. (2012) 'What do we know about urbanisation in sub-Saharan Africa and does it matter?', *International Development Planning Review* 34 (1): v–xxi.

Pratt, A. C. (2009) 'Urban regeneration: from the arts "feel good" factor to the cultural economy: a case study of Hoxton, London', *Urban Studies* 46: 1041–61.

Preston, S. A. (2009) 'The planner manager', 413–29 in Hack, G., Birth, E. L., Sedway, P. H. and Silver, J. (eds.), *Local Planning: Contemporary Problems and Prospects*. Washington, DC: ICMA Press.

Prince, R. (2012) 'Policy transfer, consultants and the geographies of governance', *Progress in Human Geography* 36: 188–203.

Pritchett, L., Woolcock, M. and Andrews, M. (2013) 'Looking like a state: techniques of persistent failure in state capability for implementation', *The Journal of Development Studies* 49: 1–18.

Qian, Z. (2010) 'Without zoning: urban development and land use controls in Houston', *Cities* 27: 31–41.

Rapoport, E. (2015) 'Sustainable urbanism in the age of Photoshop: images, experiences and the role of learning through inhabiting the international travels of a planning model', *Global Networks* 15: 307–24.

Reade, E. (1983) 'If planning is anything, maybe it can be identified', *Urban Studies* 20: 159–71.

Recio, R. B. (2021) 'How can street routines inform state regulation? Learning from informal traders in Baclaran, Metro Manila', *International Development Planning Review*, 43: 63–88.

Reimer, M., Getimis, P. and Blotevogel, H. (eds.) (2014) *Spatial Planning Systems and Practices in Europe: A Comparative Perspective on Continuity and Changes*. Abingdon: Routledge.

Reitz, J. (1998) 'How to do comparative law', *Comparative Law* 46: 617–36.

Rittel, H. W. J. and Webber, M. M. (1973) 'Dilemmas in a general theory of planning', *Policy Sciences* 4: 155–69.

Robinson, J. (2006) *Ordinary Cities: Between Modernity and Development*. Abingdon: Routledge.

Robinson, J. (2011) 'Cities in a world of cities: the comparative gesture', *International Journal of Urban and Regional Research* 35: 1–23.

Robson, B. T. (1988) *Those Inner Cities: Reconciling the Economic and Social Aims of Urban Policy*. Oxford: Oxford University Press.

Rodgers, D. T. (1998) *Atlantic Crossings: Social Politics in a Progressive Era*. Cambridge, MA: Harvard University Press.

Roggema, R. (2013) 'The design charrette', 15–34 in Roggema, R. (ed.), *The Design Charrette: Ways to Envision Sustainable Futures*. Dordrecht: Springer.

Roitman, S. and Phelps, N. (2011) 'Do gates negate the city? Gated communities' contribution to the urbanisation of suburbia in Pilar, Argentina', *Urban Studies* 48: 3487–509.

Rose, E. A. (1984) 'Philosophy and purpose of planning', 31–65 in Bruton, M. J. (ed.), *The Spirit and Purpose of Planning*, 2nd edn. London: Hutchinson.

Rose, G. (2001) *Visual Methodologies: An Introduction to the Interpretation of Visual Materials*. London: Sage.

Rose, M. H. and Mohl, R. A. (2012) *Interstate: Highway Politics and Policy since 1939*. Nashville: University of Tennessee Press.

Rose-Redwood, R. and Bigon, L. (2018) 'Gridded spaces, gridded worlds', 1–19 in Rose-Redwood, R. and Bigon, L. (eds.), *Gridded Worlds: An Urban Anthology*. Dordrecht: Springer.

Rostow, W. W. (1960). *The Stages of Economic Growth: A Non-Communist Manifesto*. Cambridge: Cambridge University Press.

Rowe, W. T. (2013) 'China: 1300–1900', 310–27 in Clark, P. (ed.), *The Oxford Handbook of Cities in World History*. Oxford: Oxford University Press.

Roy, A. (2005) 'Urban informality: toward an epistemology of planning', *Journal of the American Planning Association* 71: 147–58.

Roy, A. (2009a) 'Why India cannot plan its cities: informality, insurgence and the idiom of urbanization', *Planning Theory* 8: 76–87.

Roy, A. (2009b) 'The 21st century metropolis: new geographies of theory', *Regional Studies* 43: 819–30.

Roy, A. (2018) 'The grassroots of planning: poor people's movements, political society, and the question of rights', 143–54 in Gunder, M., Madanipour, A. and Watson, V. (eds.), *The Routledge Handbook of Planning Theory*. Abingdon: Routledge.

Roy, A. and Ong, A. (eds.) (2011) *Worlding Cities: Asian Experiments and the Art of Being Global*. Chichester: Wiley.

Rozee, L. (2014) 'A new vision for planning: there must be a better way?', *Planning Theory & Practice* 15: 124–38.

Rydin, Y. (2009) 'Sustainable construction and design in UK planning', 181–90 in Davoudi, S., Crawford, J. and Mehmood, A. (eds.), *Planning for Climate Change: Strategies for Mitigation and Adaptation for Spatial Planners*. Abingdon: Earthscan.

Rydin, Y. (2011) *The Purpose of Planning: Creating Sustainable Towns and Cities*. Bristol: Policy Press.

Rykwert, J. (2000) *The Seduction of Place: The City in the Twenty-First Century*. New York: Pantheon.

Sack, R. (2003) *A Geographical Guide to the Real and the Good*. Abingdon: Routledge.

Sager, T. Ø. (2012) *Reviving Critical Planning Theory: Dealing with Pressure, Neo-Liberalism, and Responsibility in Communicative Planning*. Abingdon: Routledge.

Salet, W. (2018) *Public Norms and Aspirations: The Turn to Institutions in Action*. Abingdon: Routledge.

Salkin, P. E. (2015) 'Land use regulation in the United States: an inter-governmental framework', 27–52 in Knaap, G.-J., Nedović-Budić, Z. and

Carbonell, A. (eds.), *Planning for States and Nation-States in the U.S. and Europe*. Cambridge, MA: Lincoln Institute of Land Policy.

Sanchez, T. and Afzalan, N. (2017) 'Mapping the knowledge domain of urban planning', 69–84 in Sanchez, T. (ed.), *Planning Knowledge and Research*. Abingdon: Routledge.

Sanyal, B. (2005a) *Comparative Planning Cultures*. Abingdon: Routledge.

Sanyal, B. (2005b) 'Hybrid planning cultures: the search for the global cultural commons', 3–28 in Sanyal, B. (ed.), *Comparative Planning Cultures*. Abingdon: Routledge.

Sargisson, L. (2018) 'Swimming against the tide: collaborative housing and practices of sharing', 145–59 in Ince, A. and Hall, S. M. (eds.), *Sharing Economies in Times of Crisis*. Abingdon: Routledge.

Sassen, S. (1991) *The Global City*. New York: Pine Forge Press.

Saunier, P. Y. (2001) 'Sketches from the urban internationale, 1910–50: voluntary associations, international institutions and US philanthropic foundations', *International Journal of Urban and Regional Research* 25: 380–403.

Saunier, P. Y. (2002) 'Taking up the bet on connections: a municipal contribution', *Contemporary European History* 11: 507–27.

Schaffer, D. (1986) 'Ideal and reality in 1930s regional planning: the case of the Tennessee Valley Authority', *Planning Perspectives* 1: 27–44.

Schön, D. A. and Rein, M. (1994) *Frame Reflection: Toward the Resolution of Intractable Policy Controversies*. New York: Basic Books.

Scitovsky, T. (1954) 'Two concepts of external economies', *Journal of Political Economy* 62: 143–51.

Scott, A. J. and Roweis, S. T. (1977) 'Urban planning in theory and practice: a reappraisal', *Environment and Planning A* 9: 1097–119.

Scott, J. (2000) *Seeing Like a State*. New Haven, CT: Yale University Press.

Sellers, J. M. (2005) 'Re-placing the nation: an agenda for comparative urban politics', *Urban Affairs Review* 40: 419–45.

Sewell, W. H., Jr (2008) 'The temporalities of capitalism', *Socio-Economic Review* 6: 517–37.

Sheller, M. and Urry, J. (2000) 'The city and the car', *International Journal of Urban and Regional Research* 24: 737–57.

Siegan, B. H. (1970) 'Non-zoning in Houston', *Journal of Law and Economics* 13: 71–147.

Sieverts, T. (2003) *Cities without Cities: An Interpretation of the Zwischenstadt*. Abingdon: Routledge.

Silva, C. N. (2015) 'Urban planning in sub-Saharan Africa: an overview', 8–40 in Silva, C. N. (ed.), *Urban Planning in Sub-Saharan Africa: Colonial and Post-Colonial Planning Cultures*. Abingdon: Routledge.

Silver, C. (2018) 'The origins of planning education: overview', 11–25 in Frank, A. I. and Silver, C. (eds.), *Urban Planning Education: Beginnings, Global Movement and Future Prospects*. Dordrecht: Springer.

Simmie, J. M. (1974) *Citizens in Conflict: The Sociology of Town Planning*. London: Hutchinson.

Skeffington Committee (1969) *People and Planning: Report of the Committee on Public Participation in Planning*. London: HMSO.

Skinner, C. and Watson, V. (2018) 'The informal economy in cities of the global south: challenges to the planning lexicon', 140–52 in Bhan, G., Srinivas, S. and Watson, V. (eds.), *The Routledge Companion to Planning in the Global South*. Abingdon: Routledge.

Skinner, G. W. (1977) *The City in Late Imperial China*. Stanford, CA: Stanford University Press.

Sklair, L. (2001) *The Transnational Capitalist Class*. Oxford: Blackwell.

Smith, D. P. and Holt, L. (2007) 'Studentification and "apprentice" gentrifiers within Britain's provincial towns and cities: extending the meaning of gentrification', *Environment and Planning A* 39: 142–61.

Smith, M. E. and Hein, C. (2018) 'The ancient past in the urban present: the use of early models in urban design', 109–20 in Hein, C. (ed.) *The Routledge Handbook of Planning History*. Abingdon: Routledge.

Smith, M. L. (2019) *Cities: The First 6,000 years*. London: Simon and Schuster.

Smith, M. P. (2017) *Reinventing Detroit: The Politics of Possibility*. Abingdon: Routledge.

Sorensen, A. (2000) 'Land readjustment and metropolitan growth: an examination of suburban land development and urban sprawl in the Tokyo metropolitan area', *Progress in Planning* 53: 217–330.

Sorensen, A. (2018) 'Planning theory and history: institutions, comparison and temporal processes', 35–45 in Hein, C. (ed.), *The Routledge Handbook of Planning History*. Abingdon: Routledge.

Stead, D. (2012) 'Best practices and policy transfer in spatial planning', *Planning Practice and Research* 27: 103–16.

Stein, S. (2019) *Capital City: Gentrification and the Real Estate State*. London: Verso.

Stern, N. (2007) *The Economics of Climate Change: The Stern Review*. Cambridge: Cambridge University Press.

Stone, D. (2004) 'Transfer agents and global networks in the "transnationalization" of policy', *Journal of European Public Policy* 11: 545–66.

Stone, J. (2014) 'Continuity and change in urban transport policy: politics, institutions and actors in Melbourne and Vancouver since 1970', *Planning Practice and Research* 29: 388–404.

Stretton, H. (1975) *Ideas for Australian Cities*. Sydney: Transit.

Subbaraman, R., O'Brien, J., Shitole, T., Shitole, S., Sawant, K., Bloom, D. E. and Patil-Deshmukh, A. (2012) 'Off the map: the health and social implications of being a non-notified slum in India', *Environment and Urbanization* 24: 643–63.

Sutcliffe, A. (1981) *Towards the Planned City: Germany, Britain, the United States and France, 1780–1914.* Oxford: Blackwell.

Swyngedouw, E., Moulaert, F. and Rodriguez, A. (2002) 'Neoliberal urbanization in Europe: large-scale urban development projects and the new urban policy', *Antipode* 34: 542–77.

Talen, E. (2005) *New Urbanism and American Planning: Conflicts of Cultures.* Abingdon: Routledge.

Talen, E. (2011) *City Rules: How Regulations Affect Urban Form.* Washington, DC: Island Press.

Tarazona Vento, A. (2015) 'Santiago Calatrava and the "power of faith": global imaginaries in Valencia', *International Journal of Urban and Regional Research* 39: 550–67.

Tarazona Vento, A. (2017) 'Mega-project meltdown: post-politics, neoliberal urban regeneration and Valencia's fiscal crisis', *Urban Studies* 54: 68–84.

Tarazona Vento, A. (2018) 'Madrid: the making of a global city region and the role of the suburbs', 18–40 in Phelps, N. A. (ed.), *Old Europe, New Suburbanization? Governance, Land, and Infrastructure in European Suburbanization.* Toronto: University of Toronto Press.

Taylor, J. E. (2002) 'The Bund: littoral space of empire in the treaty ports of East Asia', *Social History* 27: 125–42.

Taylor, P. J. (2012) 'Transition towns and world cities: towards green networks of cities', *Local Environment* 17: 495–508.

Taylor, P. J., Catalano, G. and Walker, D. R. (2002) 'Exploratory analysis of the world city network', *Urban Studies* 39: 2377–94.

Tewdwr-Jones, M. (2011) *Urban Reflections: Narratives of Place, Planning and Change.* Bristol: Policy Press.

Tewdwr-Jones, M. (2017) 'Health, cities and planning: using universities to achieve place innovation', *Perspectives in Public Health* 137: 31–4.

Tewdwr-Jones, M. and Allmendinger, P. (eds.) (2006) *Territory, Identity and Spatial Planning: Spatial Governance in a Fragmented Nation.* Abingdon: Routledge.

Tewdwr-Jones, M., Sookhoo, D. and Freestone, R. (2019) 'From Geddes' city museum to Farrell's urban room: past, present, and future at the Newcastle City Futures exhibition', *Planning Perspectives* 35: 277–97.

Thaler, R. H. and Sunstein, C. R. (2008) *Nudge: Improving Decisions about Health, Wealth, and Happiness.* New Haven, CT: Yale University Press.

Thomas, H. (2000) *Race and Planning: The UK Experience.* London: UCL Press.

Thompson, M. M. (2017) 'Citizen science as a research approach', 226–40 in Sanchez, T. (ed.), *Planning Knowledge and Research.* Abingdon: Routledge.

Thompson-Fawcett, M. (1998) 'Leon Krier and the organic revival within urban policy and practice', *Planning Perspectives* 13: 167–94.

Thrift, N. and Olds, K. (1996) 'Refiguring the economic in economic geography', *Progress in Human Geography* 20: 311–37.

Tiesdell, S. and Allmendinger, P. (2005) 'Planning tools and markets: towards an extended conceptualization', 56–76 in Adams, D., Watkins, C. and White, M. (eds.), *Planning, Public Policy and Property Markets*. Oxford: Blackwell.

Tilly, C. (1984) *Big Structures, Large Processes, Huge Comparisons*. New York: Russell Sage Foundation.

Tilly, C. (1989) 'Entanglements of European cities and states', 1–27 in Tilly, C. and Blockmans, W. P. (eds.), *Cities and the Rise of States in Europe., A.D. 1000 to 1800*. Boulder, CO: Westview Press.

Tolson, S. (2011) 'Competitions as a component of design-led development (place) procurement', 159–81 in Tiesdell, S. and Adams, D. (eds.), *Urban Design in the Real Estate Development Process*. Oxford: Wiley-Blackwell.

Turner, J. F. C. (1976) *Housing by People: Towards Autonomy in Building Environments*. London: Marion Boyars.

United Nations (1996) *Report of the United Nations Conference on Human Settlements (Habitat II)*. Istanbul: United Nations Conference on Human Settlements (Habitat II).

United Nations (2014) *World Urbanization Prospects Report*. New York: United Nations.

United Nations Human Settlements Programme (2009) *Planning Sustainable Cities: Policy Directions: Global Report on Human Settlements 2009*. New York: UN-Habitat.

Urry, J. (1990) *The Tourist Gaze: Leisure and Travel in Contemporary Societies*. London: Sage.

Urry, J. (2008) 'Governance, flows, and the end of the car system?', *Global Environmental Change* 18: 343–9.

Valler, D. and Phelps, N. (2016) 'Delivering growth? Evaluating economic governance in England's South East subregions', *Town Planning Review* 87: 5–30.

Valler, D. and Phelps, N. A. (2018) 'Framing the future: on local planning cultures and legacies', *Planning Theory & Practice* 19: 698–716.

Vasudevan, A. (2015) *Metropolitan Preoccupations: The Spatial Politics of Squatting in Berlin*. Chichester: Wiley.

Vergara-Perucich, J. F. and Arias-Loyola, M. (2019) 'Bread for advancing the right to the city: academia, grassroots groups and the first cooperative bakery in a Chilean informal settlement', *Environment and Urbanization* 31: 533–51.

Vertovec, S. (2007) 'Super-diversity and its implications', *Ethnic and Racial Studies* 30: 1024- 54.

Vickerman, R., Spiekermann, K. and Wegener, M. (1999) 'Accessibility and economic development in Europe', *Regional Studies* 33: 1–15.

Wackernagel, M., Lin, D., Evans, M., Hanscom, L. and Raven, P. (2019) 'Defying the footprint oracle: implications of country resource trends', *Sustainability* 11(2164).

Wade, R. (1990) *Governing the Market*. Princeton, NJ: Princeton University Press.

Wagner, W. (2018) 'The failure of planning in a fragmented property market: Poland's model of suburbanization', 41–65 in Phelps, N. A. (ed.), *Old Europe, New Suburbanization? Governance, Land, and Infrastructure in European Suburbanization*. Toronto: University of Toronto Press.

Ward, K. (2006) '"Policies in motion", urban management and state restructuring: the trans-local expansion of business improvement districts', *International Journal of Urban and Regional Research* 30: 54–75.

Ward, P. M. (2012) 'Self-help housing ideas and provision in the Americas', 283–310 in Sanyal, B., Vale, L. J. and Rosen, C. D. (eds.), *Planning Ideas that Matter*. Cambridge, MA: MIT Press.

Ward, S. V. (2003) 'Learning from the US: the Americanisation of western urban planning', 83–106 in Nasr, J. and Volait, M. (eds.), *Urbanism: Imported or Exported?* Chichester: Wiley.

Ward, S. V. (2005) 'A pioneer "global intelligence corps"? The internationalisation of planning practice, 1890–1939', *Town Planning Review* 76: 119–41.

Ward, S. V. (2010) 'Transnational planners in a postcolonial world', 71–96 in Healey, P. and Upton, R. (eds.), *Crossing Borders: International Exchange and Planning Practices*. Abingdon: Routledge.

Ward, S. V., Freestone, R. and Silver, C. (2011) 'The "new" planning history: reflections, issues and directions', *Town Planning Review* 82: 231–61.

Watson, V. (2009) 'Seeing from the South: refocusing urban planning on the globe's central urban issues', *Urban Studies* 46: 2259–75.

Watson, V. (2014) 'Co-production and collaboration in planning: the difference', *Planning Theory and Practice* 15: 62–76.

Webber, M. (1963) 'Order in diversity; community without propinquity', 23–56 in Wingo, L. (ed.), *Cities and Space: The Future Use of Urban Land*. Baltimore: Johns Hopkins University Press.

Webster, C. (2002) 'Property rights and the public realm: gates, greenbelts, and Gemeinschaft', *Environment and Planning B* 29: 397–412.

Wheeler, S. (2009) 'California's climate change planning: policy innovation and structural hurdles', 125–35 in Davoudi, S., Crawford, J. and Mehmood, A. (eds.), *Planning for Climate Change: Strategies for Mitigation and Adaptation for Spatial Planners*. Abingdon: Earthscan.

White, P. and Hurdley, L. (2003) 'International migration and the housing market: Japanese corporate movers in London', *Urban Studies* 40: 687–706.

White, R. and Williams, C. (2018) 'Crisis, capitalism, and the anarcho-geographies of community self-help', 175–91 in Ince, A. and Hall, S. M. (eds.), *Sharing Economies in Times of Crisis*. Abingdon: Routledge.

Whitley, R. (2007) *Business Systems and Organizational Capabilities: The Institutional Structuring of Competitive Competences*. Oxford: Oxford University Press.

Wildavsky, A. (1971) 'Does planning work?', *The Public Interest* 24: 95–104.

Wildavsky, A. (1973) 'If planning is everything, maybe it's nothing', *Policy Sciences* 4: 127–53.

Williams, C. C. (1996) 'Local exchange and trading systems: a new source of work and credit for the poor and unemployed?', *Environment and Planning A* 28: 1395–415.

Williams, J. (2019) 'Circular cities', *Urban Studies* 56: 2746–62.

Williams, R. (1986) 'Introduction', 89–90 in Masser, I. and Williams, R. (eds.), *Learning from Other Countries*. Norwich: GeoBooks.

Wilmsen, B. (2016) 'After the deluge: a longitudinal study of resettlement at the Three Gorges Dam, China', *World Development* 84: 41–54.

Wilmsen, B., Webber, M. and Duan, Y. (2011) 'Development for whom? Rural to urban resettlement at the Three Gorges Dam, China', *Asian Studies Review* 35: 21–42.

Wilson, E. (2009) 'Use of scenarios for climate change adaptation in spatial planning', 223–35 in Davoudi, S., Crawford, J. and Mehmood, A. (eds.), *Planning for Climate Change: Strategies for Mitigation and Adaptation for Spatial Planners*. Abingdon: Earthscan.

Wolman, H. (2008) 'Comparing local government systems across countries: conceptual and methodological challenges to building a field of comparative local government studies', *Environment and Planning C* 26: 87–103.

Woolcock, M. (1998) 'Social capital and economic development: toward a theoretical synthesis and policy framework', *Theory and Society* 27: 151–208.

World Bank (2013) *Cities Alliance for Cities without Slums: Action Plan for Moving Slum Upgrading to Scale*. Washington, DC: World Bank.

World Commission on Environment and Development (1987) *Our Common Future*. New York: Butterworth-Heinemann.

World Health Organization and UN-Habitat (2010) *Hidden Cities: Unmasking and Overcoming Health Inequities in Urban Settings*. Geneva: World Health Organization.

Wray, I. (2016) *Great British Plans: Who Made Them and How They Worked*. Abingdon: Routledge.

Wu, F. (2004) 'Transplanting cityscapes: the use of imagined globalization in housing commodification in Beijing', *Area* 36: 227–34.

Wu, F. (2015) *Planning for Growth*. Abingdon: Routledge.

Yiftachel, O. (1998) 'Planning and social control: exploring the dark side', *Journal of Planning Literature* 12: 395–406.

Yokohari, M., Takeuchi, K., Watanabe, T. and Yokota, S. (2008) 'Beyond greenbelts and zoning: a new planning concept for the environment of Asian mega-cities', 783–96 in Marzluff, J. M. et al. (eds.), *Urban Ecology*. Boston, MA: Springer.

Zhang, J. and Wu, F. (2006) 'China's changing economic governance: administrative annexation and the reorganization of local governments in the Yangtze River Delta', *Regional Studies* 40: 3–21.

Index